Making Sense of Genes

What are genes? What do genes do? These seemingly simple questions are in fact challenging to answer accurately. As a result, there are widespread misunderstandings and over-simplistic answers, which lead to common conceptions widely portrayed in the media, such as the existence of a gene 'for' a particular characteristic or disease. In reality, the DNA we inherit interacts continuously with the environment and functions differently as we age. What our parents hand down to us is just the beginning of our life story. This comprehensive book analyses and explains the gene concept, combining philosophical, historical, psychological and educational perspectives with current research in genetics and genomics. It summarises what we currently know and do not know about genes and the potential impact of genetics on all our lives. *Making Sense of Genes* is an accessible but rigorous introduction to contemporary genetics concepts for non-experts, undergraduate students, teachers and healthcare professionals.

KOSTAS KAMPOURAKIS is a researcher in science education at University of Geneva, where he also teaches the course Biology and Society to biology undergraduates, and various science education classes to teachers and doctoral students. He is the Editor-in-Chief of the journal *Science and Education*, the co-editor of *Newton's Apple and Other Myths about Science* (with Ronald L. Numbers, 2015), and the editor of *The Philosophy of Biology: A Companion for Educators* (2013). His book *Understanding Evolution* (Cambridge, 2014) was selected as a 2015 *Choice* Outstanding Academic Title.

Making Sense of Genes

KOSTAS KAMPOURAKIS

University of Geneva

CAMBRIDGE
UNIVERSITY PRESS

University Printing House, Cambridge CB2 8BS, United Kingdom

One Liberty Plaza, 20th Floor, New York, NY 10006, USA

477 Williamstown Road, Port Melbourne, VIC 3207, Australia

314-321, 3rd Floor, Plot 3, Splendor Forum, Jasola District Centre, New Delhi - 110025, India

79 Anson Road, #06-04/06, Singapore 079906

Cambridge University Press is part of the University of Cambridge.

It furthers the University's mission by disseminating knowledge in the pursuit of education, learning and research at the highest international levels of excellence.

www.cambridge.org
Information on this title: www.cambridge.org/9781107128132
DOI: 10.1017/9781316422939

© Kostas Kampourakis 2017

First published 2017

A catalogue record for this publication is available from the British Library

Library of Congress Cataloging in Publication data
Names: Kampourakis, Kostas, author.
Title: Making sense of genes / Kostas Kampourakis, University of Geneva.
Description: New York, NY: Cambridge University Press, 2017. |
Includes bibliographical references.
Identifiers: LCCN 2016052788| ISBN 9781107128132 (hardback) |
ISBN 9781107567498 (pbk.)
Subjects: LCSH: Genes.
Classification: LCC QH447.K36 2017 | DDC 572.8/6–dc23
LC record available at https://lccn.loc.gov/2016052788

ISBN 978-1-107-12813-2 Hardback
ISBN 978-1-107-56749-8 Paperback

To my brother, Yannis, and our mother, Evaggelia,
who have always made me think hard about
"nature" and "nurture"

Contents

Acknowledgments

There are many people I would like to thank because they made writing this book possible in various ways.

I am indebted to Bruno J. Strasser, thanks to whom I have been working at the University of Geneva for the past few years, on projects relevant to the teaching and the public understanding of genetics. His support and friendship have made writing this book possible. I am also very grateful to Andreas Müller and Didier Picard who have been supporting my work and research at the University of Geneva. My interest in human genetics goes back in time when as an MSc student I had the opportunity to work at the laboratory of Emmanouil Kanavakis at the University of Athens. I thank him for that opportunity.

While writing this book, I have been very fortunate to benefit from the thoughtful feedback of several scholars: Garland Allen, John Avise, Sheldon Krimsky, Alessandro Minelli, David S. Moore, Staffan Müller-Wille, John Parrington, Giorgos Patrinos, Erik Peterson, Gregory Radick, and Tobias Uller. I thank them all for their valuable comments and suggestions. Writing this book has also benefited from discussions during an older collaboration with Richard Burian. Of course, as is always the case, I am responsible for any remaining problems or errors.

I am also indebted to Katrina Halliday, my editor at Cambridge University Press, who supported the project from the very first moment and all along the way toward the publication of this book. Katrina has been extremely helpful and supportive, and it is a pleasure to have two books, *Understanding Evolution* and the present one, published by Cambridge University Press. Of course, besides Katrina, I am grateful to a number of people at the Press who worked with me until the publication of my book, especially Lindsey Tate, who worked with me during the production of the book and patiently dealt with all my requests. I would also like to thank Tim Oliver for designing the cover and the staff at Newgen for the copy-editing and typesetting.

Finally, there are always those to whom I owe a lot: my family. I dedicated *Understanding Evolution* to my wife Katerina and our children Mirka and Giorgos. An earlier – edited – book was dedicated to my father, Giorgos, so it is now the turn of my brother and our mother to get their own dedication. This book is most appropriate for this purpose as the striking differences and similarities among us have always made me think hard about "nature" and "nurture."

Prolegomena: Genes, Science, and Science Fiction

If one looks at mass media headlines, one will find several accounts of how genes determine various aspects of our lives. Many of these claim to take into account conclusions from recent research in genetics. The general impression is that there exist "genes for" characters,[1] i.e. that single genes cause even complex characters. This view seems to be quite prevalent e.g. it is common to find teachers teaching that genes determine characters, media reports presenting studies that found associations between particular genes and particular diseases, and personal observations of the development of characters that do not seem to be affected by the environment (Moore, 2008). A quick search on the World Wide Web reveals several examples. For instance, a 2014 article in the *Guardian* was titled "Happy gene' may increase chances of romantic relationships."[2] The title of a 2015 article in the *New York Times* suggested that "Infidelity lurks in your genes."[3] A 2014 article in *Time* magazine was titled: "The genes responsible for deadly prostate cancer discovered."[4] And there are more. Several authors have argued that messages like these impose genetic determinist views on the public (e.g. Hubbard & Wald, 1997; Nelkin & Lindee, 2004). This certainly seems plausible, particularly as many people might just read the headlines such as those mentioned previously, without ever reading the full article that might suggest otherwise. Therefore, they might conclude that genes determine who we are.

The problem of making sense of genes, i.e. understanding what genes are and what they do, has concerned me a lot and for a long time. However,

---

[1] To avoid inconsistencies while referring to features, traits, characteristics, and so on interchangeably, I am using the term "character" throughout this book, which can be defined as any recognizable feature of an organism that can exist in a variety of character states, and at several levels from the molecular to the organismal (based on Arthur, 2004, p. 212). Disease conditions will be considered as character states that deviate from what we tend to consider as "normal."

[2] www.theguardian.com/science/2014/nov/20/happy-gene-romantic-relationship-serotonin-romance

[3] www.nytimes.com/2015/05/24/opinion/sunday/infidelity-lurks-in-your-genes .html?partner=rss&emc=rss

[4] http://time.com/96247/scientists-have-discovered-the-two-genes-responsible-for-aggressive-prostate-cancer/

in my previous book, *Understanding Evolution* (Kampourakis, 2014), I refrained from using the term gene at all. Instead, I referred to genetic material and DNA sequences that are implicated in biological phenomena. Eventually, it was possible to write a whole book without any reference to genes. Yet, ignoring the problem does not contribute anything to its solution, and so I decided to devote my second book to the gene concept that was put aside in my first one. There are two reasons for this. On the one hand, the term exists in the public discourse and so it is better to try to clarify it rather than just ignore it. On the other hand, scientists use the term in their work and in its public presentation. Therefore, I thought that I could make a minor contribution to countering the public distortions of the gene concept and help students in the life sciences, biologists, biology teachers, health professionals, and anyone else interested in acquiring a better understanding of it, as well as provide them with conceptual tools to explain genes to nonexperts.

Generally speaking, our knowledge takes the form of concepts that are mental representations of the world. Concepts should be distinguished from conceptions, the latter being the different meanings of, or meanings associated with, particular concepts. This means that whereas we may generally agree on a general definition of a certain concept, e.g. "dog," people all over the world may hold different conceptions of what a dog is or looks like. In other words, even if a concept is well defined and even if it is clear to people to what this concept refers, individual conceptions may vary a lot if one takes the time to consider them. This is also the case for scientific concepts, such as the gene. Scientific concepts are systematic mental representations of the world through which explanations of and predictions about phenomena are possible (Nersessian, 2008, p. 186). In this case, the difference between concepts and conceptions becomes more striking; whereas scientists may agree on the definition of a certain concept, nonexperts may hold very different conceptions of it for various reasons. Such reasons may include the public distortions of the concept under discussion, or that people simply failed to understand it because of their own preconceptions. In the present book I focus on the gene concept that most people have heard of, but many fail to understand. My aim is to explain this concept and address certain prevalent but inaccurate conceptions. At the end of this book, the reader should have acquired a better understanding of what a gene is and is not, as well as what a gene can and cannot do.

You have probably had some genetics classes during your secondary school years. Even if you have forgotten most of what you learned at that time, you probably remember Gregor Mendel (1822–1884) and his experiments with peas. Because of these experiments Mendel is considered as a pioneer of genetics and as the person who discovered the laws of heredity. You may also remember that, according to your secondary school genetics, things were rather simple and straightforward with inherited characters. In the case of disease, for instance, your teacher used to explain to you that most "normal"[5] alleles (the different versions of the same gene) were dominant, i.e. imposed their effects on the pathogenic ones that were therefore recessive. Thus, when an individual had one normal, dominant allele and one pathogenic, recessive allele, there was no problem. However, in some cases it was possible for such individuals to have offspring with the disease, because two recessive pathogenic alleles had come together in the same individual. Why was that? According to your teacher, the normal allele somehow determined a normal character, whereas the pathogenic allele determined a pathogenic version of the same character that occurred because e.g. some important factor was missing. Thus, a person with two pathogenic alleles totally lacked that factor and so had the disease. Genes could thus determine characters and diseases.

So, what was a gene according to your high school genetics? It was usually defined as a segment of DNA that contained the information for the production (or not) of some protein[6] that in turn somehow determined a character. A definition could not be clearer, could it? In definitions like these, the take-home message is usually that genes work in a deterministic if-you-have-the-gene-you-will-also-have-the-character kind of way (Moore, 2013a).[7] However, if one looks more closely at such definitions, one will realize that genes are conceived as simultaneously operating at two levels: the molecular (production of a protein) and the organismal (determination of a character). What is implied is that the molecular level (DNA/gene) somehow determines the organismal. Even

---

[5] Defining what is normal and what is not normal is quite difficult and subjective sometimes. In this book, I will use the term in a rather vague sense to refer to whatever state can be considered as natural and unproblematic.

[6] DNA, or deoxyribonucleic acid, is a long molecule that consists of consecutive nucleotides. Proteins, or polypeptides, are long molecules that consist of consecutive amino acids.

[7] This is the topic of Chapters 6 and 7.

when gene definitions do not simultaneously refer to both levels, one may find distinct definitions referring to the molecular level and the organismal level to co-exist.

Consider, for example, the definitions of gene in two very good textbooks that I used when I was still teaching biology at school. The first of these (Walpole et al., 2011) contained the following definitions in the main text: "A gene is a particular section of a DNA strand that, when transcribed and translated, forms a specific polypeptide" (p. 67), and a "Gene [is] a heritable factor that controls a specific characteristic, or a section of DNA that codes for the formation of a polypeptide" (p. 68). In both cases, the gene is described as a section of DNA, but in the second definition it is also described as a factor that controls a character. However, if one looks at the glossary of the same book, no reference to DNA is made. A gene is defined there as "a heritable factor that controls a specific characteristic" (p. 586). The case is similar in the other textbook (Sadava et al., 2011). In the main text, genes are defined as segments or sequences of DNA that encode proteins: "Genes are specific segments of DNA encoding the information the cell uses to make proteins" (p. 6); "The sequences of DNA that encode specific proteins are transcribed into RNA and are called genes" (p. 64); and "a gene is a sequence of DNA that resides at a particular site on a chromosome, called a locus (plural loci). Genes are expressed in the phenotype mostly as proteins with particular functions, such as enzymes" (p. 242). However, the definition in the glossary of the same book pays more attention to function, ignoring structure. The gene is defined there as: "A unit of heredity. Used here as the unit of genetic function which carries the information for a single polypeptide or RNA" (p. G-12).

Is there a consistency problem here? Why are the definitions in the main text and the glossary of the same textbook different? The definitions of the term "gene" in the glossaries of both textbooks refer to a hereditary factor without specifying what exactly this is made of. In contrast, the definitions in the main text of both textbooks are specific about what is referred to by the term "gene": a section, segment, or sequence of DNA. Are these textbooks referring to the same concept at different levels of organization? In the present book, I show that the definitions in the main texts and in the glossaries of these textbooks are in fact very different. I also explain that the reasons why such different definitions co-exist in the same textbook are not pedagogical but historical. By presenting how the gene concept was coined and has evolved over the past 100 years or so, during which time research on heredity has been conducted, I show that different gene concepts have dominated discourse on heredity over different periods and that, recently, more than one have co-existed.

The next question that arises is this: What is it that genes do? If you open a newspaper or a popular magazine it is very likely that you will read a report about a recent discovery of a "gene for" something. Genes have been reported to determine characters of all kinds, such as eye color and height. They have also been reported to determine well-studied diseases, such as thalassemia and phenylketonuria, but also more complex and less-well understood ones such as coronary heart disease and cancer. Most interestingly, genes are often reported in the popular press to determine all kinds of behaviors and psychological states. Thus, "genes for" depression, schizophrenia, intelligence, alcoholism, criminality, promiscuity, homosexuality, and more have been reported to exist. As a result, genes are perceived as determining everything. This is particularly evident in characters that run in families, which are, often without a second thought, attributed to genes inherited from parents to offspring, and not to other possible factors such as their shared environment.

I speculate that if there was a report that George H. W. Bush (1924–) and his son George W. Bush (1946–) were both elected presidents of the United States because of a particular gene they both had, perhaps a "gene for" US presidency, many people would not question such a conclusion. Similarly, many people might find reasonable that there exists a "gene for" becoming a Hollywood star in the case of Kirk Douglas (1916–) and his son Michael (1944–), or in the case of Judy Garland (1922–1969) and her daughter Liza Minelli (1946–). These same people might attribute to a "gene for" the Nobel Prize the fact that both Arthur Kornberg (1918–2007) and his son Roger (1947–) were awarded a Nobel Prize – but perhaps different versions of that gene could account for the fact that Arthur's prize was in physiology and medicine, whereas Roger's was in chemistry. These examples might sound exaggerated, but as I show later in this book, claims like these are quite common in the public sphere. For many people, the interesting question is not whether genes determine characters and behaviors; the common assumption is that they do. The interesting question is how they do it.

The metaphors currently used about genes present them as autonomous entities, which both contain all the necessary information to determine characters and are capable of making use of it. Therefore, both in research and in popular parlance, genes have been described as the "essences" of life, as the absolute "determinants" of characters and disease and therefore as providing the ultimate explanations for all biological phenomena because the latter can be "reduced" to the gene level and thus be explained. These views have been described as *genetic essentialism*, *genetic determinism*, and *genetic reductionism*, respectively. They are all related to one another, and they may even seem to overlap. However, they are distinct and should not be confused. In order to avoid confusion and overlaps in definitions, in this book I use the following definitions (based on Beckwith, 2002; Kitcher, 2003; Wilkins, 2013):

- *Genetic essentialism:* genes are fixed entities, which are transferred unchanged across generations and which are the essence of what we are by specifying characters from which their existence can be inferred.
- *Genetic determinism:* genes invariably determine characters, so that the outcomes are just a little, or not at all, affected by changes in the environment or by the different environments in which individuals live.
- *Genetic reductionism:* genes provide the ultimate explanation for characters, and so the best approach to explain these is by studying phenomena at the level of genes.

Most importantly, these are the onerous conceptions that the present book aims at addressing.

Whether or not these conceptions are distinct apparently depends on how one defines them. I use these definitions in order to distinguish between three important properties usually attributed to genes: (1) that they are fixed essences; (2) that they alone determine characters notwithstanding the environment; and (3) that they best explain the presence of characters. The power attributed to genes has often gone beyond the realm of science to reach that of science fiction. Genes have been described as autonomous, self-replicating entities capable of doing everything and of determining everything. There are "fat" genes, "smart" genes, "cancer" genes, "infidelity" genes, "aggression" genes, "happiness" genes, "God" genes, and more (a World Wide Web search of these terms is illuminating;

in some cases, even books with titles like these exist). The underlying assumption in most cases is that much of what we are or do is driven (if not dictated) by our genes. Perhaps we find attributing whatever happens to one's genes very intuitive, because it makes sense immediately? It is the supernatural powers attributed to genes that this book aims at addressing. Of course, I am not going to argue that genes are not important – they are! But it is one thing to say that genes are important for what we are or do, and another that they are the ultimate determinants of these. I hope that, at the end this book, I will have succeeded at clarifying what genes are and are not, as well as what they can and cannot do.

Chapters 1–4 provide a brief account of how the initially "empty," or, to be more precise, referentially indefinite (i.e. that did not refer to a particular entity), gene concept came to have two distinct meanings during the twentieth century: that of a hypothetical inherited factor, the changes in which were somehow related to changes in characters, and that of a DNA sequence that encoded the information for a protein. Whereas it may have initially seemed self-evident that these two gene concepts might overlap and that they would converge to the same segments of DNA, by the 1970s it became quite evident that this is not the case. More recent research has shown that it is impossible to structurally individuate genes, and that the best we can do is to identify them on the basis of their functional products. I must note that in these chapters I do not intend to provide a detailed and complete history of the "gene" concept (for such histories see Beurton et al., 2000; Falk 2009; Rheinberger et al., 2015; Rheinberger & Müller-Wille, in press). Rather, these chapters aim at providing an idea of the complexities of precisely defining what a gene is.

Then, in Chapters 5–8, I describe the presentations of genes in the media and on the websites of companies selling genetic tests. I show that the underlying message in many cases is that there are genes that determine characters and disease. I also present research on the conceptions that students and the public hold about genes and the difficulties they face in understanding what genes are and do. Then, I show that simple, causal connections between genes and characters or genes and disease are not adequate to accurately represent the actual phenomena. Research in genetics shows that these are actually very complicated. In many cases, single genes cannot explain the variation observed not only for complex characters and disease but also for simple monogenic ones. On the basis of these, I conclude by explaining that genes do not actually do anything on their own. I also explain why the notion of "genes for," in the vernacular sense, is not only misleading but also entirely inaccurate and scientifically illegitimate.

Finally, in Chapters 9–12, I come to some major conclusions from the research presented in the previous chapters. First, I show that genes "operate" in the context of developmental processes only. This means that genes are implicated in the development of characters but do not determine them. Second, I explain why single genes do not alone produce characters or disease but contribute to their variation. This means that genes can account for variation in characters but cannot alone explain their origin. Third, I show that genes are not the masters of the game but are subject to complex regulatory processes. There seem to exist many regulatory sequences in what until recently has been called "junk" DNA. As a result, the genome of an organism is more than the sum of its genes. Finally, I discuss in some detail the limitations of genetic testing that are not often taken into account in public discourse, in order to show what is and what is not currently possible to achieve from DNA analyses, and to debunk the myth of their infallibility. I also show how misleading information about genes can be when it comes to probabilistic thinking.

The chapters of this book could be read independently from one another; however, in many cases individual chapters build on knowledge and understanding of concepts that have been presented in previous ones. Therefore, I recommend that you read this book from beginning to end, without skipping any chapters – unless you are very well familiar with the respective topics. However, for those readers who decide not to do so, the book includes a glossary with the definitions of the most important concepts. Next to that, there is also a guide to further reading that includes relevant books that treat in more detail many of the topics presented in this book. In most cases, I have read and cited the original research articles. However, in several cases I found the accounts given in certain books useful or the ideas illuminating, and so I am citing these. Many of the topics I present are discussed in several books, but I am only citing them wherever it is really useful. The *Further Reading* section provides information about the books one should read after reading the present one.

A central feature of the present book is that it is mostly about human characters and disease. When this is not the case, it is usually about phenomena of relevance to human life. I must note that this is not due to any anthropocentricism on my part. Quite the contrary, I believe that we are not anything special in this world, or at least that we are not any more special than any other organism that lives in it. Nevertheless, I thought that the book would be more interesting to readers if I discussed phenomena about, or relevant to, human life. I made the decision to focus on human genes because my experience as a teacher and an educator was that students' interest was aroused whenever a topic about human life or health came up. Pragmatically thinking, making sense of genes also has an important medical interest; therefore, I wanted this book to be useful not only for biologists but also for physicians and other health professionals. This approach is biased, of course, because it overlooks important aspects of life on Earth. But it is also more interesting for humans. I do hope that readers will appreciate both this decision and the outcome. I hope that they will find this biased-toward-humans book interesting and didactic. But they should also keep its bias in mind and avoid unwarranted generalizations from the mostly medical-centered and human-focused research presented in this book.

I must also note that the term "genetics" is used throughout the book in a very broad sense to refer to any research about genes. Therefore, the term "genetics" encompasses research in classical genetics of the first half of the twentieth century, molecular biology and genetics of the latter half of the twentieth century, and genomics of the past twenty-five years or so, despite the important differences among these research approaches. Conceptually, "genetics" could be perceived to refer to genes only, whereas "genomics" could be perceived to refer to the genome as a whole, including genes and everything else in DNA. Therefore, "genomics" could be considered as a broader term than "genetics," as the genome is a broader concept than the gene (see Annas & Elias, 2015, p. 3). Nevertheless, as genes have been the main focus of research so far, it is conceptually sound and certainly simple for the purpose of the present book to use the term "genetics" to refer to all research about, or relevant to, genes – no matter how these are defined – also encompassing genomics research.

The present book is intended primarily for non-experts, i.e. people not working on genetics, who want an introduction to genes. The intended audience includes undergraduate students in biology, medicine, and pharmacy, as well as biology teachers and educators. The book provides an overview of the core concepts and issues in genetics, and it can also serve as an introduction to more detailed and advanced forays in the literature. Physicians and other healthcare professionals who are interested in getting a concise overview of contemporary genetics research and concepts would also find this book useful, while it could also be useful to researchers in biology who are interested in the relationship between biological research and society at large. Finally, the book is appropriate for any lay reader who wants an accessible but rigorous introduction to genes.

I should note that the present book focuses on a research area that advances at an extremely fast pace. I found myself revising and updating the text several times during the one year or so that it took me to write this book. Therefore, I am certain that as soon as the book will be published there will be new research articles that I might have considered, should they have been published before that. However, I think that the main points of the present book and its conceptual foundations will remain unchanged for several years to come, even if we get to know more and understand the respective phenomena in more detail.

A note of caution. Throughout the book, I have used expressions such as "research has shown" or "evidence suggests," etc. Research produces data that becomes evidence within a certain theoretical framework. Neither research, nor evidence "show" or "suggest" anything on their own. Everything in science is a matter of interpretation. Nevertheless, because it would be strange for me to write each time that "research findings have formed the basis for the conclusion that..." or that "evidence has been interpreted as showing that...", I have used expressions like "research has shown" or "evidence suggests" as a kind of shorthand. Nevertheless, you should be aware that these statements have problems and which is the actual meaning behind them (I discuss this topic in more detail in Chapter 12).

Before we proceed, some nomenclature is necessary. Most genes mentioned in this book are human ones. Both gene and protein symbols in humans are written with uppercase letters. However, gene symbols are written with inclined characters, whereas protein symbols are written with regular characters. For example, the symbol for the protein called "monoamine oxidase A" is MAOA, whereas the symbol for the respective gene would be *MAOA*. For the purpose of consistency, all gene symbols and gene names in this book are derived from the Human Genome Organization Gene Nomenclature Committee website: www.genenames.org. For the few other species to the genes of which I refer in this book, I simply use the nomenclature used in the respective articles.

Let us now start trying to make sense of genes.

I Mendel and the Origins of the "Gene" Concept

If you open any biology textbook and look for information about Gregor Mendel (Figure 1.1), you will likely read that he is mentioned as an important figure in the history of biology. You will also likely read that he is considered as the most heroic figure of all. Mendel is, perhaps along with Isaac Newton (1643–1727), the exemplar instantiation of the lonely genius – a man working in isolation and discovering the laws of nature. As textbook accounts often explain, Mendel discovered the laws of heredity but did not receive the recognition he deserved during his lifetime. Contrary to other important figures in the history of biology, the narrative continues, Mendel died in oblivion because his contemporaries failed to understand the significance of his work, which was published in an obscure journal and was thus not widely read. Textbooks emphasize that it was Mendel's approach that paved the way for the science that we now call genetics. His conclusions were based on solid experimental evidence and careful mathematical analysis, but his contemporaries failed to understand his pioneering work. Mendel was a lonely pioneer of genetics, being ahead of his time – or so the story goes.

Here are some examples of such accounts from recent editions of some widely used biology textbooks: "The study of genetics which is the science of heredity, began with Mendel, who is regarded as the father of genetics" (Biggs et al., 2009, p. 277); "The modern science of genetics was started by a monk named Gregor Mendel" (Miller & Levine, 2010, p. 262); "The ground work for much of our understanding of genetics was established in the middle of the 1800s by an Austrian monk named Gregor Mendel" (Nowicki, 2012, p. 167); "Modern genetics had its genesis in an abbey garden, where a monk named Gregor Mendel documented a particulate mechanism for inheritance (Reece et al., 2012, p. 262); "Gregor Mendel is considered to be the founding father of genetics" (Ward et al., 2008, p. 98); "For the next 35 years, his [Mendel's] paper was effectively ignored yet, as scientists later discovered, it contained the entire basis of modern genetics" (Walpole et al.,

FIGURE 1.1 Gregor Mendel (© Time Life Pictures).

2011, p. 86). Note the common assumption in all textbooks: Mendel was doing genetics. In several cases, Mendel is described as the person who discovered that heredity is particulate in nature, and that the factors controlling it are those we now call genes.

The same story is also found on the World Wide Web. For instance, the Wikipedia entry about Mendel goes like this:[1] "He [Mendel] published his work in 1866, demonstrating the actions of invisible 'factors' – now called genes – in providing for visible traits in predictable ways. The profound significance of Mendel's work was not recognized until the turn of the 20th century." The relevant entry in *Encyclopedia Britannica* provides the following account of Mendel's life and work: "From the precise mathematical 3:1 ratio ... he deduced not only the existence of discrete hereditary units (genes) but also that the units were present in pairs in the pea plant and that the pairs separated during gamete formation" (Winchester, 2013). A similar account is given in the education website of the prestigious scientific journal *Nature*: "Mendel's insight

---

[1] http://en.wikipedia.org/wiki/GregorMendel.

greatly expanded the understanding of genetic inheritance ... Mendel ... hypothesized that each parent contributes some particulate matter to the offspring. He called this heritable substance 'elementen' ... Indeed, for each of the traits he examined, Mendel focused on how the elementen that determined that trait was distributed among progeny" (Miko, 2008).

Mendel, the story continues, discovered that characters are controlled by hereditary factors, the inheritance of which follows two laws: the law of segregation and the law of independent assortment. In the first case, when two plants that differ in one character, e.g. plants having seeds that are either round or wrinkled, are crossed, their offspring (generation 1) resemble one of the two parents (in this case they have round seeds). In generation 2 (the offspring of the offspring) there is a constant ratio 3:1 between the round and the wrinkled character (Figure 1.2). Round shape is controlled by factor R that is dominant, whereas wrinkled shape is controlled by factor r that is recessive. Dominant and recessive practically means that when R and r are together, it is R that dominates over r and so the respective "R" phenotype is produced as if r was not even there. This means that plants with factors RR or Rr will have round seeds, whereas plants with rr will have wrinkled seeds. The explanation of these results is that the factors (R/r) controlling the different characters (round/wrinkled) are separated (segregated) during fertilization and recombined in the offspring. This is described as Mendel's law of segregation.

When Mendel simultaneously studied the inheritance of two characters, e.g. both the shape of the seed and its color, he observed a similar but more complicated picture. When he crossed plants with yellow/round seeds and plants with green/wrinkled seeds, in generation 1, all offspring had yellow/round seeds. However, when those plants were crossed with each other, a constant ratio of 9 yellow/round: 3 yellow/wrinkled: 3 green/round: 1 green/wrinkled emerged in generation 2. This is actually the result of the combination of the probabilities to have all possible combinations of two characters, e.g. the one described earlier and a similar one regarding color.[2] Plants with factors YY or Yy have yellow seeds, whereas plants with yy have green seeds. The results

---

[2] This means that the ratio 9:3:3:1 for two characters results from the combination of the 3:1 ratio for each character ($\frac{3}{4}$ x $\frac{3}{4}$ = 9/16; $\frac{3}{4}$ x $\frac{1}{4}$ = 3/16; $\frac{1}{4}$ x $\frac{1}{4}$ = 1/16).

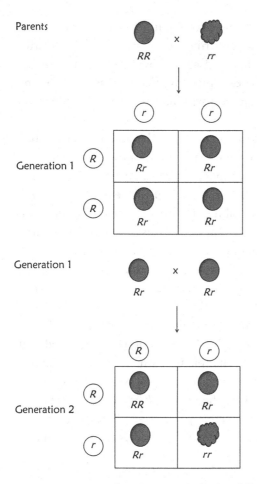

FIGURE 1.2 A cross between two plants that differ in the shape of seeds (round or wrinkled). Plants with round seeds have factors *RR* or *Rr*, whereas plants with wrinkled seeds have factors *rr*. The "wrinkled" character "disappears" in generation 1 and "reappears" in generation 2 (green peas appear here as having a darker color than yellow peas; note also that all possible combinations of gametes are made).

suggested that the factors (*R/r* and *Y/y*) controlling the different characters (seed shape and seed color, respectively) were assorted independently during fertilization. As a result, all possible combinations were obtained (yellow/round, yellow/wrinkled, green/round, green/wrinkled), and this is why these are observed in generation 2 (see Figure 1.3). This is described as Mendel's law of independent assortment.

The account that presents Mendel as the founding father of genetics, who understood that inheritance was particulate in nature, who

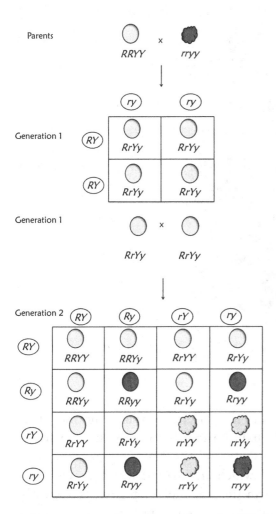

FIGURE 1.3 A cross between two plants that differ in the shape of seeds (round/wrinkled) and the color of seeds (yellow/green). In the second generation we find the characteristic ratio 9:3:3:1 (green peas appear here as having a darker color than yellow peas; note also that all possible combinations of gametes are made).

discovered the laws of heredity, who was ignored by his contemporaries, and whose reputation was established posthumously, in 1900, with the rediscovery of his pioneering paper, is quite prevalent, although it has been critically questioned since at least 1979 (Brannigan, 1979; Olby, 1979). If one looks closer into the historical details, it becomes clear that Mendel intended to study hybridization in particular and not heredity in general, whereas his 1866 paper,

FIGURE 1.4 The stereotypical image of Mendel, working alone in his garden
(© Bettmann).

titled "Versuche über Pflanzen-Hybriden" [Experiments on Plant Hybrids], does not clearly indicate that he was thinking in terms of hereditary particles. Furthermore, the laws of segregation and of independent assortment, which are attributed to him, certainly do not appear in his paper in the way biology textbooks currently describe them. In fact, many of the accounts presented in textbooks are distortions of history. In order to understand Mendel's work and contribution, we should consider it in its actual historical context. Paying attention to the details of history is important because stories about "fathers" of disciplines, such as Mendel, give the false impression that a discipline may emerge from nothing and develop independently of the surrounding context. This is most exemplarily portrayed by the usual illustration of Mendel working alone in his garden (Figure 1.4).

However, it seems that stories like these have an appeal. A recent book that is intended to provide a historical account of the development of the gene concept and that became a best seller as soon as it was published (Mukherjee, 2016) provides a triumphalist account of Mendel's life and work. According to this account, Mendel discovered genes: "We begin with Mendel's pea-flower garden, in an obscure Moravian monastery in 1864, where the 'gene' is discovered and then quickly forgotten" (p. 13). It is implied that the reason for this was that Mendel's paper was "an obscure paper ... published in a scarcely read journal" (p. 46). Mendel is also presented as having the intention to study heredity and reveal its secrets: "To reveal the nature of heredity, Mendel knew that he needed to breed hybrids" (p. 49). Mendel is also presented to have discovered that unitary hereditary factors, with dominant and recessive alleles, existed: "Had Mendel stopped his experiments here, he would already have made a major contribution to a theory of heredity. The existence of dominant and recessive alleles for a trait contradicted nineteenth-century theories of blending inheritance; the hybrids that Mendel had generated did not possess intermediate features. Only one allele had asserted itself in the hybrid, forcing the other variant trait to vanish" (p. 51); "A 'hybrid' organism, Mendel realized, was actually a *composite* – with a visible, dominant allele and a latent, recessive allele" (p. 51); "Each trait was *unitary* – distinct, separate, and indelible. Mendel did not give this unit of heredity a name, but he had discovered the most essential features of a gene" (p. 53). But, unfortunately, Mendel's contemporaries failed to understand the importance of his work: "That three papers in the short span of three months in 1900 independently converged on Mendel's work was a demonstration of the sustained myopia of biologists, who had ignored his work for nearly forty years" (p. 60).[3] Mendel is, therefore, presented in Mukherjee (2016) as a lonely pioneer of genetics who was ahead of his time. The stereotype of the lonely genius is emphasized in that book too. But is this what the history of science suggests? Let us now take a closer look into this history.

Mendel was born in Moravia, which was a province of the Austrian Empire and is now part of the Czech Republic. Moravia was, at that time, a world-leading center of breeding practices (Wood & Orel, 2005). One such practice was hybridization, during which offspring were produced

---

[3] Not surprisingly, neither Olby (1985) nor Olby (1966) is included in the "selected bibliography" of Mukherjee (2016). If you want to read a rich and concise history of the gene concept, and of the roles that it played within science, then Rheinberger and Müller-Wille (in press) is a must-read book for you.

from the cross-fertilization of individuals from different species.[4] An important figure in these efforts was Cyril Napp (1792–1867), the abbot of the Augustinian Monastery of St. Thomas at Brno (Brünn), who accepted Mendel there in 1843. Napp also supported Mendel's studies at Vienna, from 1851 to 1853, in physics with Christian Doppler (1803–1853) and in chemistry, paleontology, and plant physiology with Franz Unger (1800–1870). Unger had argued that all plants were descended from common ancestors, and had also proposed a theory of evolution (Gliboff, 1998). He thus rejected the idea of species fixity and argued that the plant world had developed gradually. Unger seems to have had a major influence on Mendel, who attended his classes in the early 1850s. It seems probable that Mendel learned from him about the hybridization experiments of Josef Gottlieb Kölreuter (1733–1806) and Carl Friedrich von Gärtner (1772–1850) (Olby, 1985, pp. 95–100). The question whether hybridization could produce new species goes back to Carl Linnaeus (1707–1778), who had published a treatise on this topic in 1751, arguing that hybridization could indeed produce new species. In contrast, both Kölreuter and Gärtner had come to the conclusion that hybridization could not produce new species. Gärtner's extensive work, published in 1849, had led to the conclusion that hybrids were not variable but constant, meaning that the same hybrid forms were produced again and again if one combined the same parent species. Kölreuter also thought that hybridization could not occur in nature but only artificially (for the details of this work and the broader cultural and intellectual context, see Roberts, 1929; Olby, 1985; Müller-Wille & Rheinberger, 2012).

Mendel returned to Brno in 1853. The question of whether hybridization could produce new species, and the broader understanding of this phenomenon, was of interest for the practical purposes of breeding related to agriculture and the socioeconomic context of Brno. Mendel's monastery was at that time a center of learning, particularly in agriculture. In addition, the practical and financial benefits of new agricultural practices were a top priority for the monastery, as most of its income came from extensive land leasing to local farmers. It was in this context

---

[4] The species concept is difficult to define (see Wilkins, 2009; Richards, 2010). I describe these difficulties in some detail in chapter 6 of Kampourakis (2014). This concept is used throughout the present book, rather loosely, to refer to a group of individuals that are reproductively isolated from other groups and/or genetically distinct. For sexually reproducing organisms, a species can be defined as a number of, usually similar, organisms that can interbreed and produce fertile offspring.

that Mendel began his hybridization experiments in 1856. He selected thirty-four distinct varieties of the edible pea (*Pisum sativum*) for his experiments, and subjected them to a two-year trial for purity, in order to obtain varieties that when self-reproduced always produced plants with the same characters. Then he performed crosses between different varieties, focusing on seven characters: the shape of the seed (round or wrinkled); the color of the seed (yellow or green); the color of the seed coat (white or gray-brown); the shape of the ripe pod (smoothly arched or deeply ridged between seeds); the color of the unripe pod (green or yellow); the position of flowers (axillary or terminal); and the length of the stem (1.9–2.2m or 0.24–0.46m) (Orel, 1984, p. 44). Mendel concluded his experiments with *Pisum* in 1863 (see Orel, 1984; Olby, 1985; Allen, 2003; Gliboff, 2013, for more details on Mendel's work).[5]

Mendel's results were presented in the meetings of the Brno Natural Science Society on February 8 and March 8, 1865, and were published in the society's journal in 1866. In the beginning of his paper Mendel expressed his aim to "to follow the development of hybrids in their descendants". He also noted that, until that time "a universally valid law describing the formation and development of hybrids has not yet been established".[6] It is important to note that Mendel was interested in studying the transmission of characters over generations bred from hybrids, and to better understand how this happened. In his paper Mendel described the transmission of characters rather than that of hereditary particles like genes. In particular, Mendel observed that the hybrids obtained from the various crosses between different varieties were not always intermediate between the parental forms. In contrast, some hybrids exhibited certain characters exactly as they appeared in the parental plants. Mendel called dominant the parental characters that appeared in the hybrids, and recessive the parental characters that did not appear in the hybrids but that reappeared fully formed in the next generation. Thus, Mendel studied and wrote about characters and not about hereditary particles, and so did not discover that heredity was particulate in nature. More generally, he studied hybridization and not

---

[5] Mendel studied intraspecific hybrids, i.e. hybrids stemming from crosses between varieties of the same species, not between different species.

[6] All quotations from Mendel's paper are from a new translation by Kersten Hall and Staffan Müller-Wille, titled "Experiments on Plant Hybrids" and available at http://centimedia.org/bshs-translations/mendel/. I am very grateful to both of them for granting me access to it before its publication.

heredity, and it should therefore be no surprise that the term "heredity" does not appear in his paper.[7] A careful study of his paper also shows that, strictly speaking, Mendel did not discover the two laws commonly attributed to him; rather, he observed their consequences under the particular experimental conditions. Statements that look like the laws of segregation and independent assortment can be found in his paper, but they are not explicitly described as such; they are also quite different from the way they are currently described in textbooks and elsewhere (see Kampourakis, 2015, for a more detailed account).[8]

The term "heredity" in the modern, biological sense, i.e. with reference to the transmission of some substance across generations, does not appear in writings on the generation of organisms until the mid-eighteenth century. The systematic use of this term was initially done in medical contexts around 1800 by French physicians, and was soon introduced to other European languages (López Beltrán, 2004, 2007; Cobb, 2006). The term "heredity" derives from the Latin *hereditas*, which means inheritance of succession. The biological concept of heredity resulted from the metaphorical use of a juridical concept, which referred to the distribution of status, property, and other goods, according to a system of rules about how these should be passed on to other people once the proprietor passed away (Müller-Wille & Rheinberger, 2012, pp. 5–6). In other words, heredity, in the modern biological sense, is a rather "recent" concept. However, it is in this sense that this term is mostly used today, whereas the term "inheritance" is used both in biological and nonbiological contexts. Therefore, in the present book, "genetic inheritance" will refer to the process of transmission of genetic material across generations, whereas "heredity" will refer to the broader phenomenon of which this process is part. In this sense, "heredity" is considered as an exclusively biological concept, whereas "inheritance" is not.

The earliest references to heredity can be found in Herbert Spencer's (1820–1903) *Principles of Biology* (1864), just a year before Mendel's paper was presented. At that time, the mechanism of heredity was

---

[7] Even though the noun *heredity* (*Vererbung*) does not exist in Mendel's paper, the verb *inherit* (*vererben*) does. Mendel also used quite frequently verbs that could be translated as "transmit" (übergehen ... auf; übertragen) and are also used in the context of inheritance (Müller-Wille, personal communication).

[8] Interestingly, Jonathan Marks (2008) has argued that Mendel's two laws were invented in the form that they are widely known by Thomas Hunt Morgan (1866–1945) in 1916.

at the center of biological thought, in part because Charles Darwin's (1809–1882) theory of descent with modification through natural selection (published in 1859 in the *Origin of Species*) lacked a complementary theory that could explain the origin and inheritance of new variations that were so central to it.[9] Darwin wrote: "The laws governing inheritance are quite unknown; no one can say why the same peculiarity in different individuals of the same species, and in individuals of different species, is sometimes inherited and sometimes not so" (Darwin, 1859, p. 13). In response to this problem, Darwin proposed in 1868 his *Provisional Hypothesis of Pangenesis*, a term that he attributed to Hippocrates (460 BCE–370 BCE), which literally means "origin from everywhere" (*pan*: all; *genesis*: origin). According to this hypothesis, all parts of the body participated in the formation of the offspring by producing microscopic entities, the gemmules, which somehow carried the organismal properties from generation to generation. Herbert Spencer had also proposed in 1864 a theory of heredity based on minute hereditary determinants, as did Ernst Haeckel (1834–1919) in 1876. Both Spencer and Haeckel accepted that the inheritance of acquired characters was possible, and this idea was also central in Darwin's *Pangenesis* (Kampourakis, 2013; Allen, 2014).

In 1871, Francis Galton (1822–1911) tried to test experimentally the hypothesis of *Pangenesis* and practically disconfirmed it (Galton, 1871a).[10] In 1876, he proposed his own theory of heredity, suggesting that hereditary factors did not arise from the various body tissues. He also proposed the term "stirp" that accounted for the total of the hereditary elements or germs at the fertilized ovum, thus postulating a form of a germline theory, i.e. that reproductive cells existed separately from body (somatic) cells. Galton also suggested that evolutionary change might take place in a discontinuous manner and not gradually. William Keith Brooks (1848–1908) took this idea further in 1883 to suggest that evolution proceeded with extensive modifications, and not gradually as Darwin had suggested. Carl von Nägeli (1817–1891) proposed a theory that combined older views (spontaneous generation) and modern views

---

[9] It should be noted that in the nineteenth century heredity was strongly associated with variation and was largely discussed as hereditary variation (see Müller-Wille & Rheinberger, 2012, chapter 5).

[10] There is an interesting exchange on this topic between Darwin and Galton in *Nature* (see Darwin, 1871; Galton, 1871b).

(the existence of hereditary factors). Hugo de Vries (1848–1935) retained some of Darwin's ideas, and by combining breeding experiments with Galton's statistical methods he suggested in 1889 a modified theory of pangenesis. In the meantime, during the 1880s, vital dyes and improved microscopes made possible the visualization of cellular structures and processes. In this way it was also shown that reproductive cells existed independently of the rest of the body tissues. August Weismann (1834–1914) drew on these new findings to propose the shift from "pangenesis" to "blastogenesis": the idea that characters were inherited only from the germline and not from the whole body.[11] This led to the abandonment of some of Darwin's ideas, including the inheritance of acquired characters that Weismann strongly rejected (Kampourakis, 2013). Some central features of these theories are presented in Table 1.1, and the respective scholars are presented in Figure 1.5.

All these people were aware of one another's work and practically formed a scientific community, actively and interactively working to develop a theory of heredity. Mendel is nowhere in this picture. Only Nägeli came to know of Mendel's experimental work, through their correspondence from 1866 until 1873. Following Nägeli's advice, Mendel worked on *Hieracium* (hawkweed, a genus of the sunflower family) from 1866 to 1871, which gave different results from those of *Pisum*. Nägeli did not seem to pay much attention to Mendel's work, yet on at least one occasion he cited Mendel's 1866 paper. Most importantly, the Brno Natural Science Society sent more than 100 copies of the journal that included Mendel's paper to scientific centers around the world. At least ten references to Mendel's paper appeared in the scientific literature before 1900, some of them in books that were widely read by naturalists. Therefore, it was possible for Mendel's work to become more widely known during his lifetime. Why did it not? Probably because it was not an explicit attempt to develop a theory of heredity that was of interest to naturalists at that time (Olby, 1985; Kampourakis, 2013). Mendel was rather interested in understanding hybridization and its

---

[11] See Weismann (1893/1892, p. xiii). Weismann's account of inheritance also gave rise to neo-Darwinism in the original sense of the term. Neo-Darwinians were "neo" because they explained adaptation exclusively on the basis of the elimination of unfit and preservation of fit variants in germline factors by natural selection, and not like Darwin himself who relied on use and disuse as an explanation of some adaptations and assumed that acquired characters could be passed on to descendants (Depew & Weber, 1995, pp. 187–191).

TABLE 1.1 *Theories of Heredity of the Latter Half of the Nineteenth Century*

Author (Publication Year)	Hereditary Factors	Mechanism for Variation	Acquired Characters Inherited
Herbert Spencer (1864)	Physiological units contained in cells	Physiological units were remolded	Yes
Charles Darwin (1868)	Gemmules thrown off from every unit of the body	Deficient amount of gemmules or modification of gemmules	Yes
Ernst Haeckel (1876)	Plastidules with a frequency and amplitude of vibration	The frequency and amplitude of the vibration of plastidules could change due to the influence of external conditions	Yes
Francis Galton (1876 & 1889)	Stirp (sum total of developed and latent germs) existing in the body	Germs were not identical and might be modified if remained latent for long	No
William Keith Brooks (1883)	Gemmules present in all cells	Gemmules, thrown off by affected cells, transmitted the change to the ovum	No
Carl von Nägeli (1884)	Determinants contained in the idioplasm	Internal perfecting forces, external stimuli causing changes to the idioplasm, and sexual reproduction could give rise to intermediate characteristics	No
Hugo de Vries (1889)	Pangens existing in cell nucleus into two groups (active and inactive)	Pangens of the same kind but of different origin might be activated or pangens might be slightly dissimilar to the original ones after cell division	No
August Weismann (1880s & 1892)	Biophors existing in the cell nucleus and forming units of higher order (determinants, ids, idants)	Modification of certain germ-plasm determinants and differential combination of parental ids during fertilization	No

Source: Based on Kampourakis (2013), except for Haeckel's theory that is described in Allen (2014).

FIGURE 1.5 Some of the naturalists and scholars who tried to develop theories of heredity during the nineteenth century: (a) Charles Darwin (© Bettmann); (b) Herbert Spencer (© Science & Society Picture Library); (c) Francis Galton (© Science & Society Picture Library); (d) William Keith Brooks (© JHU Sheridan Libraries/Gado); (e) Carl von Nägeli (reproduced from Roberts, 1929, p. 184); (f) Ernst Haeckel (© Stock Montage); (g) August Weismann (© Bettmann); (h) Hugo de Vries (© Bettmann).

patterns, which would be of practical, agricultural interest (Olby, 1985; Orel & Wood, 2000). It was in this practical, local context that Mendel's work made sense during his time.

However, important developments during the latter half of the nineteenth century would later provide a new context for reading Mendel's paper. On the one hand, as already mentioned, first Galton and then Weismann developed frameworks of "hard" heredity, i.e., one characterized by discontinuous variation and nonblending characters. Both Galton and Weismann rejected the idea of the inheritance of acquired characters; also, Galton postulated and Weismann established the idea of the germline, i.e., that reproductive cells exist independently of the other cells of the body. In the late 1870s, Walther Flemming (1843–1905) observed and described mitosis and Oscar Hertwig (1849–1922) observed and described meiosis.[12] In the 1880s, Eduard Strasburger (1844–1912) concluded that fertilization involved the fusion of two nuclei, and Edouard van Beneden (1846–1910) described how this took place at the level of chromosomes. Theodor Boveri's (1862–1915) experiments during the 1890s provided evidence for the role of the nucleus and chromosomes in heredity. Edmund Beecher Wilson (1856–1939) published a monograph titled *The Cell in Development and Inheritance* (Wilson, 1896), which was the standard textbook for a whole generation. In this book he noted that the nucleus contained the basis of heredity (for details see Bowler, 1989, pp. 74–92; Carlson, 2004, pp. 23–28).

In 1900, Hugo de Vries, Carl Correns (1864–1933), and Erich von Tschermak (1871–1962) published the results of their research on plant hybridization, which agreed with those obtained by Mendel (Correns, 1950/1900; de Vries, 1950/1900; Tschermak, 1950/1900).[13] De Vries actually published two papers on this topic, but in the first one he did not mention Mendel. Correns, a student of Nägeli who might have been

---

[12] Mitosis is the division of the cell nucleus that finally results in two cells having exactly the same number of chromosomes as each other and as the initial cell. This is how somatic cells proliferate. Meiosis is the division of the nucleus that finally results in four cells having half the number of chromosomes of the initial cell. This is how reproductive cells are formed (see Figure 3.5).

[13] One might be tempted to think here that these three people were working on the same questions and that eventually arrived at the "rediscovery" of Mendel's conclusions because the science of that time was mature enough to achieve this. However, if one looks closer into the work of Correns, de Vries, and Tschermak, one will find enormous differences. Therefore, why the simultaneous "rediscovery" took place is not simple and straightforward to explain (see Brannigan, 1981, pp. 117–119).

long aware of Mendel's work, insisted on Mendel's priority over both de Vries and himself, perhaps in an attempt to resolve a potential priority dispute. However, it seems that de Vries did not intend to overlook Mendel's work in order to claim priority. Rather, it seems that he did not really think that Mendel's work was that important (see Brannigan, 1981, pp. 90–96; Olby, 1985, pp. 109–133). Nevertheless, this simultaneous "rediscovery" brought Mendel back to the scene. Read in a new context, his paper was considered as bringing together the findings of breeding experiments and cytology, showing that particulate determinants existing in the nucleus of the cell were segregated and independently assorted. But this happened in 1900. Mendel was an outsider to the community that developed theories of heredity based on invisible hereditary factors during the latter half of the nineteenth century (summarized in Table 1.1 and presented in Figure 1.5) from which the discipline of genetics actually emerged.

Two important points should be emphasized here. The first point is that the usual presentation of Mendel as a heroic, lonely pioneer of genetics distorts the actual history of genetics and also conveys an inauthentic image of how science is actually done. The portrayal of a whole discipline emerging from the work of an isolated individual is one of the most widespread myths about science (Olesko, 2015), which masks the fact that science is a human activity, done within scientific communities, in particular, social, cultural, religious, and political contexts. The second point is that scientific questions usually arise out of economic or technological ones, rather than from human curiosity alone. Mendel carried out his experiments in the context of a series of practical questions related to agriculture. This is why he was studying hybridization and why he was not trying to develop a theory of heredity. Contrary to what many textbooks still claim, Mendel contributed virtually nothing to the development of a theory of heredity during the latter half of the nineteenth century. This does not undermine the importance of Mendel's experimental approach for genetics. But one should clearly distinguish between the impact of Mendel's experiments in the context in which he conducted them in the 1850s–1860s, and their impact in the new context in which they were reconsidered and reinterpreted in the 1900s.

2   The Genes of Classical Genetics

After 1900, the work of Mendel guided the development of the new science of "genetics," a term coined by William Bateson (1861–1926) (Figure 2.1a). Bateson first mentioned the term "genetics" in a 1905 letter to a friend, noting that for "a professorship relating to Heredity and Variation ... No single word in common use quite gives this meaning. Such a word is badly wanted and if it were desirable to coin one, 'Genetics' might do" (quoted in Dunn, 1991/1965, p. 69). The term appeared in print the next year, in a book review that Bateson wrote. He also proposed the term in 1906, during his inaugural address to the Third Conference on Hybridization and Plant Breeding of the Royal Horticultural Society. The term was adopted for the published proceedings the next year, describing the event as the Third International Conference on Genetics (Dunn, 1991/1965, pp. 68–69; Olby, 2000). So the stage for the new science was set.

Bateson's book *Mendel's Principles of Heredity: A Defence* contains the first English translation of Mendel's paper (Bateson, 1902, pp. 40–95). In this book, Bateson presented Mendel's work as providing the solutions for various problems relevant to heredity. He also introduced new terms such as "allelomorph," "heterozygote," and "homozygote." Allelomorph referred to the different versions of the same character and to the respective factors. In 1927, George Shull (1874–1954) introduced the shorter term, "allele" (Shull, 1935), which became the common term that refers to the alternative versions of the same gene. Currently, humans and other organisms are considered to carry pairs of alleles; an individual carrying the same allele twice is described as a homozygote, whereas another carrying two different alleles is described as a heterozygote. In that book Bateson also provided his own explanation for the neglect of Mendel's work: "It may seem surprising that a work of such importance should so long have failed to find recognition and to become current in the world of science. It is true that the journal in which it appeared is scarce, but this circumstance has seldom long delayed general recognition. The cause is unquestionably to be found in that neglect of the experimental study of the problem of Species which supervened on the general acceptance of the Darwinian doctrines" (p. 37).

FIGURE 2.1 (a) William Bateson, who coined the term "genetics" (© Universal Images Group); (b) Wilhelm Johannsen, who coined the term "gene" (© Paul Popper/ Popperfoto).

Bateson had been influenced by Galton and Brooks, and considered discontinuous variation as having enormous importance, certainly being more important than continuous variation that Darwin favored. For this reason, he was in debate with the biometricians Karl Pearson (1857–1936) and Raphael Weldon (1860–1906) who considered themselves as Galton's followers. Bateson criticized them in his 1902 book for resisting the importance of Mendel's work in understanding heredity. In the heart of the debate was the relative importance of continuous and discontinuous variation for evolution (see Gillham, 2001, pp. 303–323). But in the same year, Weldon showed that Mendel's "laws" might not actually work even for peas. Weldon's studies of varieties of pea hybrids led him to conclude that there was a continuum of colors from greenish yellow to yellowish green, as well as a continuum of shapes from smooth to wrinkled. It thus appeared that in obtaining purebred plants for his experiments, Mendel had actually eliminated all natural variations in peas, and that characters were not as discontinuous as he had assumed (Weldon, 1902; Jamieson & Radick, 2013). So, less than two years after the rediscovery of Mendel's work, it was questionable how generalizable his conclusions were.

William Castle (1867–1962) also reported several exceptions to Mendel's ratios in 1903. Between 1904 and 1908, Bateson and his colleagues also reported deviations from these ratios and realized that these were not universal. According to the reinterpretation of Mendel's paper, when two heterozygotes for two characters were crossed (AaBb x AaBb), the phenotypic ratio in the offspring would be 9:3:3:1. In particular, 9 out of 16 offspring would exhibit the two dominant characters, 3 out of 16 one dominant and one recessive character, 3 out of 16 the other dominant and the other recessive characters, and 1 out of 16 would bear the two recessive characters (Bateson, 1902, p. 11; see also Figure 1.3). But Bateson and his colleagues soon observed other ratios, such as 15:1 and 9:7. These were explained as the result of epistasis; some characters were produced from the contribution of two distinct factors of which one affected the other (Carlson, 2004, pp. 122–124). These new epistatic phenomena did not lead to a reconsideration of mendelism, but rather to its modification in order to accommodate the new findings. Mendel's work looked too good to be abandoned.

One reason for this was that Mendel's work helped new observations make sense, as well as produce new observations in the first place. Perhaps the most interesting immediate implication was the understanding of the role of chromosomes in heredity. In 1903, Walter Sutton (1877–1916) provided cytological evidence that explained Mendel's ratios, based on the understanding of meiosis of that time. Sutton was concerned about recent observations indicating that the maternal and the paternal chromosomes remained independent. This implied that reproductive cells should contain either the maternal or the paternal chromosomes. Sutton performed a careful study of the process of cell division and concluded that a large number of different combinations of maternal and paternal chromosomes were possible in the mature reproductive cells of an individual. After a detailed analysis, Sutton concluded: "the phenomena of germ-cell division and of heredity are seen to have the same essential features, viz., purity of units (chromosomes, characters) and the independent transmission of the same" (Sutton, 1903, p. 237). Despite a critical flaw (Hegreness & Meselson, 2007), Sutton's insight brought cytology and genetics even closer together, putting the foundations for explaining the physical basis of Mendel's ratios and for understanding the chromosomal nature of heredity.

In 1909, Wilhelm Johannsen (1857–1927) (Figure 2.1b) proposed the term "gene" to refer to the hereditary factors. Etymologically, the term derives from the hereditary factors of de Vries' (Intracellular) Pangenesis, which were called "Pangens" and were occasionally transcribed as "Pangenes," whereas the idea goes back to Darwin's *Pangenesis*. Johannsen suggested that it was only the second part of this term that should be retained.[1] He also noted that: "The word gene is completely free from any hypothesis; it only expresses the established fact, that at least many properties of an organism are conditioned by special, separable and thus independent 'conditions', 'foundations', 'dispositions' " (translated in Roll-Hansen, 2014, p. 4). For Johannsen, the gene concept was practically undefined, and free from any assumption about its localization in the cell and its material constitution. However, he considered genes as real entities and contrasted them to the speculative hereditary particles proposed in earlier theories during the previous fifty years (see Table 1.1): "The genes are realities, not hypothetical conceptions, like so many entities that have previously been presented in a purely speculative manner like Darwin's gemmules, Weismann's biophores, de Vries' pangenes, (first time 1889) etc." (translated in Roll-Hansen, 2014, p. 5). So, the first conceptualization of the gene was not accompanied by a specific hypothesis about its nature. For Johannsen the gene was "nothing but a very applicable little word" that might be "useful as an expression of the 'unit factors', 'elements' or 'allelomorphs' in the gametes, demonstrated by modern Mendelian researchers" (Johannsen, 1911, p. 132).[2] There was thus no need to be more specific about the nature of the hereditary factors.[3]

[1] As Moss (2003, p. 2) correctly points out, the etymology of the term "gene" highlights the significance of the concept: it is something out of which other things arise.

[2] Johannsen was also the one to coin the terms "genotype" and "phenotype": "A 'genotype' is the sum total of all the 'genes' in a gamete or zygote... All 'types' of organisms, distinguishable by direct inspection or only by finer methods of measuring or description, may be characterized as 'phenotypes' " (Johannsen, 1911, pp. 133–134).

[3] One reason that the idea of hereditary factors became widely accepted among biologists seems to have been the dominance of a mechanistic-materialist philosophy in the beginning of the twentieth century. According to this philosophy, each organism was composed by distinct parts to which it should be broken down in order to be studied, and experimentation was the best way to do this. It was under the influence of this philosophy that genes were conceived as the "atoms" of biology, which was thus no longer different from chemistry and physics. Furthermore, the initial applications of this new mendelian genetics were quite successful, and almost all phenomena could

FIGURE 2.2 Thomas Hunt Morgan established an influential experimental research programme for the study of genes, and made the term more widely known (© Keystone-France).

Many answers to questions about genes were given by an influential research program led by Thomas Hunt Morgan (1866–1945) (Figure 2.2) and his collaborators at Columbia University. This program also established *Drosophila* (the fruit fly) as a model organism in genetics research (for the broader context and the details of the research program, see Allen, 1978a; Kohler, 1994). As early as 1913, Morgan tried to clarify the relation between "unit characters" and "unit factors" – he had not yet adopted the term "gene": "The confusion is due to a tendency, sometimes unintentional, to speak of a unit character as the product of a particular unit factor acting alone, but this identification has no real basis" (Morgan, 1913, p. 5). Using several examples, Morgan explained that to think that one factor alone could determine a character was misleading and stemmed from a misunderstanding of the notations used. It might be a surprise for you to read that the simplistic notion of factors (genes in today's parlance) that alone determine phenotypic characters, still taught in biology classes today, was rejected by early researchers in genetics. In contrast, Morgan and the other researchers were perfectly aware that such simplistic notions did not appropriately represent biological phenomena.

be explained under that light with no or some modifications. As a result, the gene became and remained the central concept of classical genetics (see Allen, 2014).

Consider the following excerpt from a classic book written by Morgan and his students (Morgan et al., 1915, pp. 208–210):

> Mendelian heredity has taught us that the germ cells must contain many factors that affect the same character. Red eye color in Drosophila, for example, must be due to a large number of factors, for as many as 25 mutations for eye color at different loci have already come to light ... Each such color may be the product of 25 factors (probably of many more) and each set of 25 or more differs from the normal in a different factor. It is this one different factor that we regard as the "unit factor" for this particular effect, but obviously it is only one of the 25 unit factors that are producing this effect ... The converse relation is also true, namely, that a single factor may affect more than one character ... Failure to realize the importance of these two points, namely, that a single factor may have several effects, and that a single character may depend on many factors, has led to much confusion.

This is the important distinction between *genes as character makers* and genes as *character-difference makers*: it is one thing to suggest that a gene makes a character, and another that a gene makes a difference in the state of a character (this is discussed in detail in Chapter 10). Apparently, researchers more than 100 years ago had already obtained experimental evidence indicating the complexities of development. It was clear to them that there is no single *gene-character* connection, as a single gene might affect various characters and many genes might affect a single character. Beyond that, though, they understood neither how genes affected characters, nor what exactly their effects on these characters were (see also Waters, 1994).

The way Morgan and his colleagues described the role of genes seems to have served two roles: (1) it indicated that between genes and characters there is a many-to-many, and not a one-to-one relationship, and (2) that, notwithstanding this, it was possible to suggest that it is changes in particular genes that bring about changes in particular characters. This has been described as the differential concept of gene: there is a relation between a difference in a gene and a difference in a character. Two considerations are important here. The first one is that it is not entirely clear whether it is the change itself in a particular gene, or if it is the effect of the other twenty-four unchanged genes, which produce the difference in the character. The second consideration is that this account of changes in genes that produce changes in characters totally ignores environmental influences. The reason for this is that the research by Morgan and his colleagues was conducted in controlled laboratory conditions, where environmental changes did not take place and could thus be ignored. This is why the different characters that the same gene could produce in different environments were not considered until much later (Schwartz, 2000; see also Chapters 9 and 10).

Although Morgan and his colleagues considered the relationship between gene and characters as many-to-many one, one can find in the very same book the idea of a "gene for" a character. For instance, we read (Morgan et al., 1915, p. 11):

> As shown in this diagram, a spermatozoon bearing *the factor for long wings* fertilizing an egg bearing the same factor produces a fly pure for long wings; a spermatozoon bearing *the factor for long wings* fertilizing an egg bearing *the factor for vestigial*[4] *wings* produces a hybrid fly that has long wings. (emphases added)

This way of referring to "factors for characters" is used throughout that book, and we read about "the factor of ebony" (p. 13), "the factor for red" (p. 18), and so on. But early in the book the authors explained that whereas it was "customary to speak of a particular character as the product of a single factor," everyone familiar with the phenomena of Mendelian inheritance was aware that the "so-called unit character is only the most obvious or most significant product of the postulated factor" (Morgan et al., 1915, p. 32). Therefore, a certain form of a gene that brought about the character described as vestigial wings could be described as the gene *for* vestigial wings. However, it was clear to those researchers that the notion of "genes for" was used only to indicate a correlation; a mutation in gene X brings about a different version of character Y. This was briefly described as "X is the gene for Y," but it was clear to researchers what this shorthand meant. However, even today in the public sphere this shorthand makes people draw unwarranted conclusions, which are discussed in Chapter 5. It is thus very important to note that the idea of genetic determinism may have not stemmed directly from researchers themselves, even though they might have themselves used genetic determinist expressions. To understand

---

[4] Refers to flies that only have small vestiges of the wings, and not fully formed ones.

the prevalence of genetic determinist ideas, it is also important to consider their social and economic origins. Eugenics developed worldwide between 1900 and 1940, and it was especially prominent in the United States, Britain, and Germany. Whereas genetics arguments were used, there were also political motivations. One of the major arguments in all countries was that it was not efficient to let genetic defects spread in the populations and deal with the consequences of taking care of these people. This argument is clearly socioeconomical and not genetic (Allen, 1997).

Another important idea in the book by Morgan and his colleagues was that genes are located sequentially on chromosomes like beads on a string (Morgan et al., 1915, pp. 131–132) referring to a figure on p. 60 of the same book (Figure 2.3a). This figure is indicative of how genes were conceived as discrete parts of chromosomes. Morgan and his collaborators had realized that several genes were not inherited independently; in contrast there was some kind of genetic linkage between them that in turn pointed to a physical linkage. Thus, genes could be conceived as beads on the same string. This conceptualization was essential for the process of crossing over[5] depicted in that figure, and the techniques used subsequently by Morgan and his colleagues for mapping genes. The first genetic map, i.e. the first map showing the linear arrangement of genes on chromosomes had been published two years earlier by Alfred Sturtevant (1891–1970) (Figure 2.3b), a student of Morgan (Sturtevant, 1913). In that paper, Sturtevant also set out the logic for genetic mapping. The number of crossovers per 100 cases was used as an index of the distance between any two genes (still described as factors in that paper). If one could thus determine the distances between genes A and B and between genes B and C, it would also be able to predict the distance between genes A and C. Therefore, the relative positions of genes could be empirically mapped on chromosomes, and this gave them a more material character than before. This understanding became possible at that time in part because of luck. Morgan and his collaborators worked with *Drosophila* that has four chromosomes only, and this made the identification of genes that were linked more

---

[5] Crossing over is the phenomenon of exchange of chromosome parts between two homologous chromosomes during meiosis, the cell division leading to the production of reproductive cells (gametes). This phenomenon results in new combinations of genes in offspring that did not exist in their parents.

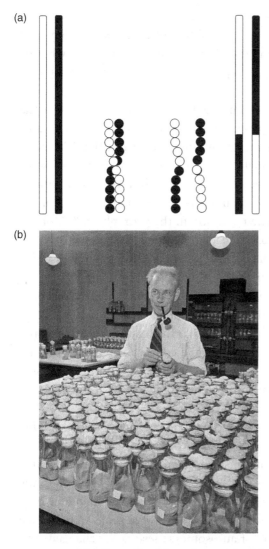

FIGURE 2.3 (a) A figure like this in the 1915 book by Morgan and his colleagues represented chromosomes like strings consisting of beads that were the genes, called unit factors at that time. It also depicted how crossing over took place and resulted in different combinations of genes, e.g. "black & white," instead of "black only" or "white only" (based on Morgan et al., 1915, p. 60). (b) Alfred Sturtevant produced the first map showing the linear arrangement of genes on chromosomes; here in front of numerous bottles with fruit-flies (© Bernard Hoffman).

probable than if the organism had forty-six chromosomes, as we do. Figure 2.4 presents how the alleles of different genes can be combined in the reproductive cells (gametes) after meiosis, depending on whether they are linked or not, for a simple case such as an individual with genotype RrYy of Figure 1.3.

Morgan and his colleagues dominated the field described as classical genetics. It was due to their adoption of the term "gene" that it became widely used. Morgan used the term for the first time in 1917: "The germ plasm must, therefore, be made up of independent elements of some kind. It is these elements that we call genetic factors or more briefly genes. This evidence teaches us nothing further about the nature of the postulated genes, or of their location in the germ plasm." (Morgan, 1917, pp. 514–515). Then Morgan made the interesting remark: "Why, it may be asked, is it not simpler to deal with the characters themselves, as in fact Mendel did, rather than introduce an imaginary entity, the gene" (p. 517). He nevertheless defended the need to use the concept of gene, even if there was no evidence about the nature and localization of genes. This was done for heuristic purposes that had to do with explaining the phenomena observed, as the gene concept had already been a very valuable heuristic tool for conducting research (for a discussion of the conceptual shift from unit factors to genes, see Darden, 1991, pp. 168–190).

Morgan and his colleagues studied several characters in *Drosophila*, such as eye color, which were related to sex chromosomes and so exhibited different ratios than those observed by Mendel and presented in Figures 1.2 and 1.3. Ratios like 3:1 and 9:3:3:1 were observed in the majority of cases for genes located on chromosomes found in both sexes, which are called autosomes.[6] However, when genes were located on sex chromosomes instead of autosomes, these ratios changed. The reason for this is that, whereas in organisms like *Drosophila* there always exist two alleles for a gene located on autosomes, there are some genes located on X chromosomes for which there is no corresponding allele in males because they only have one X chromosome. In this case, assuming that there exist two alleles, e.g. X^A and X^a, females can be homozygous (X^AX^A or X^aX^a) or heterozygous (X^AX^a), whereas males are

---

[6] Sex chromosomes were identified for the first time in 1905 by Nettie M. Stevens (1861–1912) and Edmund B. Wilson (Brush, 1978; Ogilvie & Choquette, 1981).

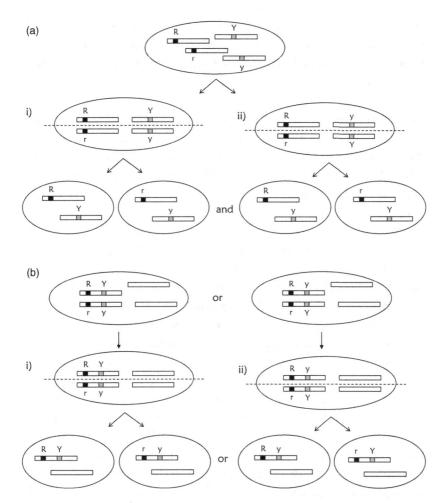

FIGURE 2.4 (a) How the alleles of the same gene are segregated and how the alleles of two genes located on two different chromosomes are independently assorted in the gametes during meiosis. In some cells procedure (i) will take place whereas in others procedure (ii) will take place, the two procedures being equally probable. Because the genes are independent, all possible gametes (RY, Ry, rY, ry) are produced by an individual. (b) When genes are located on the same chromosome, and are thus "linked," not all combinations are possible. What matters in this case is which alleles are linked (e.g. whether R is linked with y and r with Y, or R with Y and r with y). An individual will thus have either gametes RY and ry (i) or Ry and rY (ii) The replication of the genetic material is ignored in this figure, as it was not known at the time, although it was believed that genes could be replicated. Therefore, the figure should only be considered in qualitative, not quantitative, terms.

hemizygous (either X^AY or X^aY). As shown in Figure 2.5, males and females can exhibit different characters. In particular, whether it is the female parent or the male parent that has white eyes makes a difference in which characters the offspring will have. The offspring of a male fly with white eyes and a female fly with red eyes are different from the offspring of a male fly with red eyes and a female fly with white eyes. In both cases, half of the offspring will have red eyes and half of the offspring will have white eyes. However, in the first case there will be both males and females with either red or white eyes, whereas in the second case only males will have white eyes and only females will have red eyes (Figure 2.5).[7]

A major contribution summarizing the conclusions from the work of Morgan's group was his book *The Theory of the Gene* (Morgan, 1926), in which he presented all the available evidence showing that genes were located on chromosomes. According to him, the most complete and convincing evidence for the importance of chromosomes in heredity came from research that showed the specific effects of particular changes in the number of chromosomes. It had been observed that flies lacking one chromosome 4 developed to be slightly different in many parts of the body compared to the "normal" ones. These results showed that the presence of a single chromosome 4 was not adequate to produce a "normal" phenotype, and that therefore some crucial genes should be located on the missing one (Morgan, 1926, pp. 45–48). Morgan described the theory of the gene as one that was based on particular principles: (1) that the paired genes related to the characters of an individual; (2) that the genes of the same pair separated in accordance with Mendel' s first law; (3) that genes that were not linked assorted independently in accordance with Mendel's second law; (4) that crossing over took place; and (5) that crossing over provided evidence for the linear order of genes and their relative positions (p. 25). Morgan's theory was also based on certain assumptions: that genes were relatively constant, that they could be multiplied, and that they were united and then separated during the maturation of the reproductive cells (p. 27). He concluded the book with a section titled: "Are genes of the order of organic molecules?" His conclusion was that this question was difficult

---

[7] Figure 3.5 presents how the gametes of the individuals X^aY and X^AX^a in Figure 2.5 are produced during meiosis.

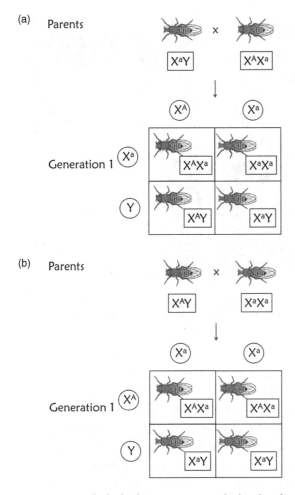

FIGURE 2.5 X-linked inheritance was studied and understood by Morgan and his colleagues. When the male has white eyes and the female has red eyes (being darker than white eyes in this figure), half of the male and half of the female offspring will have white eyes, whereas the other halves will have red eyes (a). But when it is the male that has red eyes and the female that has white eyes, all the female offspring will have red eyes and all the male offspring will have white eyes (b).

to answer. However, he estimated that genes should have the size of the larger organic molecules, and that they were constant exactly because they were chemical entities (pp. 309–310).

The next year Hermann J. Muller (1890–1967) (Figure 2.6a), one of Morgan's students and main collaborators, provided crucial evidence that genes could be chemical entities (Muller, 1927). In that article he

FIGURE 2.6 Two scientists who provided crucial evidence that genes are material entities: (a) Hermann Muller (© Bettmann) (b) Barbara McClintock (© Universal History Archive).

reported that treatment of sperm with relatively heavy doses of X-rays caused an increase in mutation rate in *Drosophila* of about 15,000 percent in the treatment group compared to the control group. What was more interesting was that the characteristics of the mutations produced by X-rays were generally similar to those previously observed in *Drosophila*. Actually, most of the already known mutant phenotypes were produced during the experiments. Muller noted that the interest of these experiments lied "in their bearing on the problems of the composition and behavior of chromosomes and genes" (p. 86). That his experiments produced most of the mutant phenotypes already observed in *Drosophila* suggested that genes had some material composition that was altered by X-rays (see Falk, 2009, pp. 131–140). This provided crucial evidence that genes should be material entities of some kind.

Additional, crucial evidence that genes were material entities came from studies of crossing over (Figure 2.3a). This phenomenon was taken for granted as early as 1915, but it was only in 1931 that Barbara McClintock (1902–1992) (Figure 2.6b) and her colleague Harriet Creighton (1909–2004) showed that there was a correspondence between chromosomes exchanging parts and the recombination of phenotypes, and therefore that genetic information and genes were carried on chromosomes. As the authors put it, their aim was to show that cytological crossing over occurs and that it is accompanied by genetic crossing over. McClintock had found that in a certain strain of corn, chromosome 9 had "a conspicuous knob at the end of the short arm." Therefore, in order to show the correlation between cytological and genetic crossing over, it was necessary to have plants that differed in the presence of the knob and of particular linked genes. By crossing plants that differed appropriately in these characteristics, it was shown that cytological crossing over occurred and that was accompanied by the expected types of genetic crossing over. It was thus concluded that chromosomes of the same pair exchanged parts at the same time they exchanged genes assumed to be located on these regions (Creighton & McClintock, 1931).

Morgan received the 1933 Nobel Prize in Physiology and Medicine "for his discoveries concerning the role played by the chromosome in heredity." In his Nobel lecture he emphasized that it would be "somewhat hazardous to apply only the simpler rules of Mendelian inheritance; for, the development of many inherited characters depends both on the presence of modifying factors and on the external environment for their expression." In other words, Morgan suggested that the simple Mendelian genetics could not account for the development of phenotypes as "the gene generally produces more than one visible effect on the individual, and … there may be also many invisible effects of the same gene." The view that genes affect but do not determine characters continued to characterize his thinking. In the same lecture Morgan also clearly explained why at that point it did not make much difference for geneticists to be aware of what genes were made of: "There is no consensus of opinion amongst geneticists as to what the genes are – whether they are real or purely fictitious – because at the level at which the genetic experiments lie, it does not make the slightest difference whether the gene is a hypothetical unit, or whether the gene is a material particle."[8] Therefore, the gene was still a hypothetical entity, a conceptual tool with a heuristic value. By that time, it had been found that genes were located on chromosomes, and thus could be made by some material substance. But the definitive answer to the question "what

[8] "Thomas H. Morgan – Nobel Lecture: The Relation of Genetics to Physiology and Medicine." Nobelprize.org. Nobel Media AB 2014. Web. www.nobelprize.org/nobelprizes/medicine/laureates/1933/morgan-lecture.html

genes were made of" made no actual difference in genetics research at that time.

How about practical applications? It is important to note that most of the geneticists of that time were interested in agricultural applications. As shown in the previous chapter, this was exactly the case with Mendel. Agriculture was used to support research in genetics and this is why the discipline developed as quickly as it did (see Sapp, 2003, pp. 137–138). Therefore, it is no surprise that the appendix in Allen (2014) of "geneticists associated with agricultural or commercial organizations and institutions or funded by them" includes Bateson, Johannsen, and several other important figures of genetics of that time. Two of them were George Shull, already mentioned earlier as the person who first used the term "allele," and Edward East (1879–1938). Shull and East were pioneers in the development of hybrid corn, one of the important advancements in genetics research, as well as in the commercialization of the products of science. When two varieties of corn A and B had specific properties, such as high yield and resistance to disease, it was possible to cross A and B individuals and produce hybrid forms H that had both of these properties. This was an amazing application of genetics as farmers could buy the superior quality hybrid seeds and produce high-yielding and disease-resistant plants. However, when hybrids were crossed, or self-crossed, they produced a variety of different plants, a lot of which were of the A and B type and did not have both of the much-wanted properties. As a result, it was more efficient for farmers to buy new H seeds each year, rather than plant the seeds emerging from the crosses of hybrids. This of course benefited the companies that created the H type in the first place, and which sold seeds to farmers every year. Therefore, the invention of hybrid corn resulted in the production of a superior quality plant, but also to a copyrighted, commercial product that yielded gains to the company that had created it (Lewontin, 1993, pp. 53–57).

Besides agriculture, another major application of the findings of classical genetics was in understanding (and controlling) human heredity. A very interesting application was the genetic analysis that John Burdon Sanderson Haldane (1892–1964) performed for the inheritance of hemophilia in the family of Queen Victoria (1819–1901).[9] In an article

---

[9] A few years earlier, following Sturtevant's principles, Haldane (1936) had published the first map of human chromosome X, showing the relative positions of genes related to several kinds of eye and skin disease.

titled, perhaps provocatively, "Blood Royal," Haldane explained that an "ultra-microscopical particle called a gene" on the X chromosome was "somehow concerned in making a substance needed for blood coagulation" (Haldane, 1939, p. 27).[10] In the case of mutation, he continued, a normal gene could become a hemophilic gene. As the respective gene was located on the X chromosome, women would have no problem, as they had two X chromosomes and thus could have one "normal" allele that was adequate for proper coagulation. But as men had only one X chromosome, they would either be "normal" or hemophilic – in the latter case with a risk to die from excess bleeding (hemophilia is inherited exactly as shown in Figure 2.5, if you substitute it for the white eye). Haldane then noted that one son, three grandsons, and six great-grandsons of Queen Victoria had been hemophiliacs, whereas her father, grandfather, and uncles all lived long enough to indicate that they did not have hemophilia. By analyzing their family tree, Haldane concluded: "The gene must have originated by mutation, and the most probable place and time where the mutation may have occurred was in the nucleus of a cell in one of the testicles of Edward, Duke of Kent, Victoria's father, in the year 1818" (p. 30).[11] Haldane finished that article with an even more important recommendation: "There is a great deal to be said for eugenics. I trust that I have shown that it is not only among the 'lower' classes that eugenic principles might be applied" (p. 31). With this, Haldane referred to the discrimination in the application of eugenics principles and politics that were popular at the time.

Although Galton was not the first to make eugenic proposals, he was the one who coined the term "eugenics," literally meaning "good birth." He was very interested in interventions that would increase the proportion of favorable characters and decrease the proportion of unfavorable ones in human populations (Waller, 2001; Renwick, 2011). Such policies were widespread in various European countries and the United States during the first half of the twentieth century, and their implementation was characterized by biases and discrimination. Relevant policies and laws were passed in various countries, which could be roughly described

---

[10] This article was initially published as Haldane (1938).

[11] The type of hemophilia (A or B) and its molecular origin had never been identified, as there are no known living descendants of Queen Victoria who carry the disease. However, the discovery and the identification of the presumed remains of the family of Czar Nicolas II have shown that it was type B hemophilia (Lannoy & Hermans, 2010).

as being of two types. On the one hand, "negative eugenics" aimed at discouraging the "less fit" or inferior members of society from having children, and this was achieved even through their sterilization. On the other hand, "positive eugenics" was about encouraging in various ways the "fittest" or most superior individuals to reproduce early and often. After the monstrosities of World War II, a new field emerged that was freed for these problems, which is currently known as human genetics (for a detailed history of eugenics see Kevles, 1995; Paul, 1995, 1998; for an accessible overview see Paul, 2014; for the history of the emergence of genetic medicine from eugenics see Lindee, 2005; Comfort, 2012).

From all the above, it is concluded that a particular gene concept, that of a hypothetical entity whose nature and structure was entirely unknown, characterized research in and practical applications of genetics during the first half of the twentieth century. In the beginning, genes could only be studied through breeding experiments. Therefore, they were initially conceptualized as hypothetical entities, which had a heuristic value as they could be used to account for the phenotypic differences observed among organisms. This gene concept, which I will call the *classical gene* hereafter, proved an extremely useful tool for doing research, even though its explanatory potential was relatively limited. Toward the end of the period described as classical genetics, genes started to take a more material form, being considered as physical objects that were linearly arranged on chromosomes and that were each related to some character. However, their nature was still unknown. It was the understanding of the molecular nature of genes that took place in the 1940s and especially in the 1950s that brought about a new conception of genes. That was the molecular gene concept, the nature of which was – in contrast to the gene concept of classical genetics – specified in a definite manner. This happened through a shift in research interests and approaches that began during the period of classical genetics and brought about a new molecular vision of life.

3   The Molecularization of Genes

Genes were conceptualized in more material and more concrete terms as a better understanding of biochemical and molecular processes emerged. Until the 1940s, the dominant approach had been to study the effect of mutations on organisms and work out the details of the respective biochemical mechanisms. However, in 1941 George Beadle (1903–1989), a former student of Morgan at Columbia University, and his colleague Edward Tatum (1909–1975) came up with a new idea: instead of understanding the biochemistry of gene products, they thought that they might start with well-known metabolic compounds and work out their genetics (Kohler, 1994, p. 233). To do this with *Drosophila* would of course be difficult because it is a rather complex organism for this kind of study. Therefore, Beadle and Tatum turned to a much simpler organism, the bread mold *Neurospora crassa*. In a 1941 paper, they published their results, concluding that a single gene seemed to somehow correspond to a single protein in a certain metabolic pathway (Beadle & Tatum, 1941). They found that the inability in mutant *Neurospora* strains to synthesize particular molecules, such as vitamins B1 and B6, was "inherited as though differentiated from normal by single genes"[1] (Beadle & Tatum, 1941, p. 506). What they managed was to show that mutant *Neurospora* strains, requiring one or the other vitamin to survive, were the outcome of mutations in single genes. This in turn showed that genes somehow regulated biochemical processes and the synthesis of enzymes.

This relation became widely known as the "one gene-one enzyme hypothesis." Norman Horowitz (1915–2005) coined the phrase to describe the achievement of his collaborators, and used it as the title of a paper he presented at the 1948 meeting of the Genetics Society of America (Horowitz, 1948). In that paper, he argued that it was possible to calculate, based on experimental data, the frequency of genes with a single function to be at least 0.74 (74 percent). But this "one gene-one enzyme" concept was in marked contrast with the one advanced

---

[1] Once more genes conceived as character-difference makers (more on this in Chapter 10).

by Morgan and his colleagues, which has already been described in Chapter 2. Whereas geneticists accepted the many-to-many relationship between genes and characters, Beadle and Tatum concluded that, at least at the biochemical level, there were one-to-one relationships. Several scientists, such as Max Delbrück (1906–1981), were skeptical because they thought that several genes controlled the synthesis of an enzyme (Morange, 1998, pp. 27–28). Nevertheless, an important conceptual advance was brought about by Beadle and Tatum: the relation between genotype and phenotype could thereafter be conceptualized in terms of gene and gene product. This was very important because, as discussed in Chapter 2, geneticists of the classical era could only describe the effects of changes in genes on phenotypic characters. After the work of Beadle and Tatum, it was possible to describe the effects of genes in more concrete, molecular terms. Yet, what genes were made of remained unknown. Interestingly, as the work of Beadle and Tatum brought closer the concepts of gene and of proteins as gene products, there were many scientists who were tempted to think that genes and proteins were one and the same thing (Olby, 1994, p. 152).

The landmark year for the elucidation of the nature and structure of genes is considered to be 1953 (for the relevant history, see Olby, 1994; Judson, 1996). James Watson (1928–) and Francis Crick (1916–2004) were young and ambitious when they became interested in the structure of DNA (Figure 3.1a). Linus Pauling (1901–1994), a prominent chemist of the time, was also working on the same question. Pauling was on constant competition in the field of discovering the structure of complex molecules with William Lawrence Bragg (1890–1971) who was the head of the Cavendish laboratory at Cambridge where Watson and Crick worked. Bragg would thus likely be happy for them to work on the structure of DNA and find it before Pauling. Watson and Crick performed no experiments for this purpose. They adopted Pauling's model-building approach in order to figure out the structure of DNA, and relied on experimental data provided by others. Erwin Chargaff (1905–2002) had shown that any DNA molecule contained equal proportions of adenine and thymine, as well as of guanine and cytosine. John Griffith (1928–1972) had pointed out that adenine and thymine, as well as guanine and cytosine, could fit together, linked by hydrogen bonds. Maurice Wilkins (1916–2004) and Rosalind Franklin (1920–1958) had performed X-ray diffraction studies of DNA, and their results

FIGURE 3.1 (a) James Watson (left) and Francis Crick (right) insightfully combined different kinds of data and came up with the double helix model of DNA (© Bettmann); (b) The actual model of the DNA double helix that Watson and Crick constructed in 1953 (© Science & Society Picture Library).

suggested a spiral structure for the molecule. Watson and Crick built actual models on the basis of this data that helped them arrive at the double helix structure (Figure 3.1b) of DNA (see chapters 19–22 of Olby, 1994; chapter 1 of Judson, 1996).[2]

On the basis of all this, in April 1953 Watson and Crick famously proposed the double helix model (Figure 3.2) for the structure of DNA (Watson & Crick, 1953a). The same volume of *Nature* also included two papers that provided the evidence for this model. One of these was coauthored by Wilkins, in which the evidence from X-ray diffraction was analyzed to show that the structure was indeed helical and that it existed in a natural state (Wilkins et al., 1953). The other paper was coauthored by Franklin, and provided further support for the "probably helical" structure model suggested by Watson and Crick (Franklin & Gosling, 1953) (see also Figure 3.3). Many people may think that the "discovery" of the double helix had an immediate impact and revolutionized research in molecular biology. The model was there; the crystallographic evidence was also there – so the revolution of molecular biology had begun? No, it was not as simple, immediate, and straightforward as that.

But let us first see the details of the double helix model of DNA. According to this, a DNA molecule consists of two nucleotide strands linked to each other. Each of these strands in turn consists of nucleotides, which differ only in their constitution of each of four

---

[2] There are various accounts and views about how exactly Watson and Crick acknowledged (or not) the contributions of these people, and more generally whether Franklin, Chargaff, and others received the recognition they deserved – Wilkins shared the Nobel Prize with Watson and Crick in 1962. The details fall outside the scope of the book, but it must be noted that the model that Watson and Crick proposed clearly stemmed from the contributions of other people. Watson's account in his book *The Double Helix* (Watson, 1996) made many people think of Franklin as a victim of discrimination because she was a woman, and therefore whose contributions to this huge achievement were not appropriately acknowledged. However, several women working with Franklin at King's argued that there was no discrimination of any such kind. In addition, the results of her work that Watson and Crick got without her knowing it were included in a report of the Medical Research Council, which was not confidential and which was distributed to several people. Among them was Max Perutz, who got it and gave it to Crick. Wilkins had also shown Watson and Crick a high quality photograph that Franklin had taken and that indicated crucial dimensions for the structure. Thus, Crick and Watson insightfully combined the crucial data that Franklin and Gosling had found with Chargaff's discovery of the 1:1 ratio of the bases to develop the double helix model of DNA (Olby, 1994, pp. 385–423; Judson, 1996, pp. 619–637).

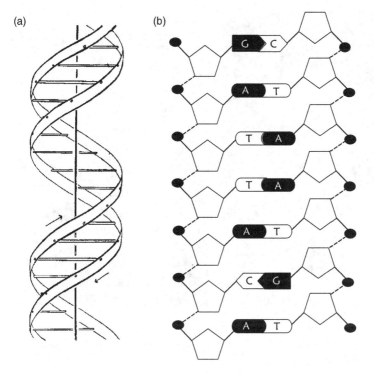

FIGURE 3.2 (a) The double helix structure of DNA as imagined by Watson and Crick (1953a, p. 737) and as drawn by Crick's wife Odile for their *Nature* paper (see Olby, 2009, pp. 184–185). Reprinted by permission from Macmillan Publishers Ltd: Nature (pp. 171, 737–738), © 1953. (b) A representation of the structure of a DNA molecule. All nucleotides have the same basic structure and differ in which of the four bases A, T, G, or C, they contain. Note that A and G are similar, and so are T and C. Also, A is complementary to T, whereas G is complementary to C. The resulting DNA molecule consists of two complementary and antiparallel nucleotide strands.

"bases": adenine (A), thymine (T), cytosine (C), and guanine (G). The nucleotides of the same strand are connected with relatively strong bonds in a linear sequence. The nucleotides of the opposite strands are connected through relatively weak bonds that are formed always between an A and a T or a C and a G. There exist two bonds within each A-T pair and three bonds within each G-C pair, so although they are weak they altogether contribute to a rather s-like structure (Figure 3.2). In their article, Watson and Crick noted, "the specific pairing we have postulated immediately suggests a possible copying mechanism for the genetic material" (Watson & Crick, 1953a, p. 737). This simply meant

FIGURE 3.3 Two scientists who provided crucial crystallographic evidence for the structure of DNA: (a) Rosalind Franklin (© Universal History Archive); (b) Maurice Wilkins (© Bettmann).

that, as there were only two possible pairs of nucleotides (A-T and G-C), the two strands would be complementary, i.e. if there was a certain base at a certain point of one strand, then the other base of the A-T/G-C pair should be on the respective point on the other strand. For example, if the sequence in one strand were GATTACA, then it could be easily inferred that the sequence in the respective position in the other strand would be CTAATGT (because A pairs with T, and G with C).

In a subsequent article, Watson and Crick suggested that their proposed structure might help solve a fundamental biological problem: "the molecular basis for the template needed for genetic replication. The hypothesis we are suggesting is that the template is the pattern of bases formed by one chain of the deoxyribonucleic acid and that the gene contains a complementary pair of such templates" (Watson & Crick, 1953b, p. 966). Watson and Crick wrote at the beginning of this second article that the importance of DNA for cells could not be disputed, as many lines of evidence suggested that it was "the carrier of a part of (if not all) the genetic specificity of the chromosome and thus of the gene itself" (Watson & Crick, 1953b, p. 964). Almost a decade earlier, Oswald Avery (1877–1955), Colin MacLeod (1909–1972), and Maclyn McCarty (1911–2005) had published experimental results suggesting that DNA was the hereditary material. In their experiments it was found that DNA was the material that transformed bacteria of one type to another (Avery et al., 1944). This conclusion was not widely accepted, because at the time many believed that only proteins, not DNA, had the necessary specificity to be genes. One reason for this is that proteins consist of twenty different amino acids, whereas DNA consists of four different nucleotides. Therefore, for a molecule that consists of 100 subunits, there can be 20^{100} different combinations if it is a protein and 4^{100}, a lot less, if it is DNA. Proteins were thus considered as more variable than DNA, and therefore as more capable of having the necessary specificity to be the genetic material. Eventually, in 1952, Alfred Hershey (1908–1997) and Martha Chase (1927–2003) conclusively showed that most of the material from a bacteriophage (a virus infecting bacteria) entering a bacterium was nucleic acid and not protein (Hershey & Chase, 1952). This made many researchers accept that DNA was the genetic material.

In their second 1953 article, Watson and Crick also admitted: "Until now, however, no evidence has been presented to show how it might carry out the essential operation required of a genetic material, that of exact self-duplication" (Watson & Crick, 1953b, p. 965). So, initially, not much attention was given to the double helix, and the proposition of the double helix model did not cause a revolution all at once (Olby, 2003).[3] Significant support for the double helical structure of DNA became available only a few years later when Matthew Meselson (1930–) and Franklin Stahl (1929–) showed that each of the two new DNA molecules produced during DNA replication consisted of one strand from the initial DNA molecule and a newly synthesized one. This suggested that when a DNA molecule was replicated, each strand served as a template for the synthesis of a complementary one (Meselson & Stahl, 1958). For

---

[3] There was some publicity after Watson, Crick, and Wilkins were awarded the Nobel Prize in 1962, but it was only in the 1990s that numerous references to the Watson and Crick 1953 papers (1953a, 1953b) appeared. The reason seems to have been the beginning of the Human Genome Project: the attempt to sequence the human genome and the expectations it raised (Strasser, 2003; see also Chapter 9).

example, a DNA molecule consisting of an GTGT and a CACA strand would produce two molecules after replication. One of them would consist of the initial GTGT strand and a new CACA one. The other molecule would consist of the initial CACA strand and a new GTGT one (Figure 3.4a). It should be noted that the term "self-duplication" used by Watson and Crick in their article, as quoted earlier, and the general description of the process of DNA replication as "self-replication", has supported reference to DNA as a self-replicating molecule. However, DNA neither replicates itself nor is intrinsically stable. The process of DNA replication depends on various proteins, such as DNA helicases and DNA polymerases (Figure 3.4b), needed for opening and copying the molecule, as well as for correcting various kinds of mistakes occurring during replication or due to mutagens (Keller, 2000, pp. 26–27; Hubbard, 2013; for details see Griffiths et al., 2005, chapter 7). DNA replication takes place before cell division so that the genetic material in the cell-to-divide is doubled. In this way, the new cells will have the same number of chromosomes with the initial cell after mitosis, and half that number after meiosis. A healthy human male can produce enough sperm cells to release about 20,000,000 of these in each ejaculation. As shown in Figure 3.5, approximately half of these sperm cells will carry chromosome X, whereas the other half will carry chromosome Y. The possibilities for an ovum to be fertilized by a sperm cell that carries either the X or the Y chromosome is thus approximately 50 percent. The presence of the Y chromosome affects whether the offspring will be male or female (see Chapter 9 for more details).

We currently know that mistakes during DNA replication are possible. One example is the change from a G-C pair to an A-T one. The base G normally occurs in the "keto" form (H–N–C=O); however, it can occasionally change to the less stable "enol" form (N=C–OH). The latter form of G pairs with T rather than with C. As a result, during DNA replication it is a T and not a C that is inserted opposite an enol-G. At the next round of replication T will be paired with A, thus producing a permanent heritable change in the sequence of DNA, as the pair in that position will thereafter be A-T and not G-C, as it was in the initial molecule. A point mutation will have thus occurred. Mutations can also occur because of mutagens, i.e. factors directly affecting the structure of DNA. One example is that of UV light that causes adjacent Ts to join. When these are later excised and repaired, there is a possibility

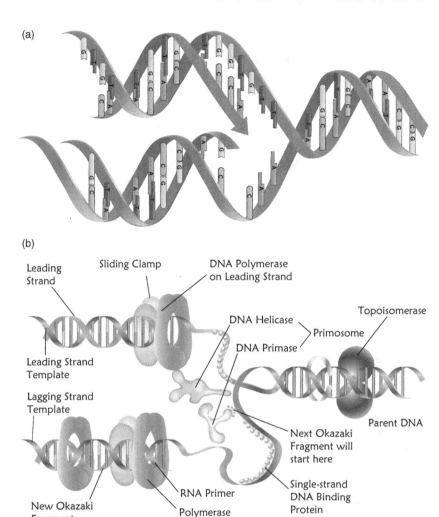

(a)

(b)

Leading Strand

Sliding Clamp

DNA Polymerase on Leading Strand

Leading Strand Template

Lagging Strand Template

DNA Helicase

DNA Primase

Primosome

Topoisomerase

Parent DNA

Next Okazaki Fragment will start here

Single-strand DNA Binding Protein

New Okazaki Fragment

RNA Primer

Polymerase

FIGURE 3.4 (a) An overview of DNA replication, in which each strand of the initial DNA molecule serves as a template for the synthesis of the complementary strand. Thus, two new molecules emerge, each consisting of an old and a newly synthesized strand. These two new molecules would be identical if there were no mutation errors. This is the portrayal of DNA as a self-replicating molecule, i.e. which can replicate on its own and produce two identical molecules (© ellepigrafica). (b) A more detailed and more accurate representation of the process of DNA replication, which includes some of the various proteins participating in the process. This portrayal shows that DNA is not a self-replicating molecule because it depends on various proteins for its replication. This is very important as DNA replication can take place only when these proteins are present (either in vivo, i.e. within cells, or in vitro, i.e. in the test tube as e.g. during the polymerase chain reaction) (© snapgalleria).

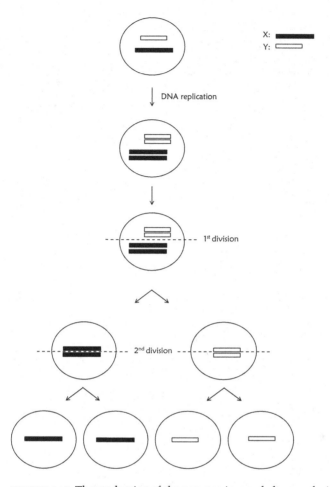

FIGURE 3.5 The production of the gametes in a male human during meiosis. Approximately half of the resulting sperm cells will have the X chromosome and the other half will have the Y chromosome.

of error as other bases might take their place (Slack, 2014, pp. 36–37). Such changes can produce variations in DNA sequences that may or may not affect a gene product in some way. These variations are usually described as mutations when they bring about a significant phenotypic change and as polymorphisms when no such change takes place. Generally speaking, we can distinguish between germline mutations, i.e. those that occur in any germ (reproductive) cell of the body and that we usually inherit from our parents, and somatic mutations, i.e.

those that occur in any cell of our body (except for the germ cells) after conception, such as those that initiate the development of cancers.

It was during the 1950s that the first evidence that mutations in genes affecting the structure of proteins are related to disease became available. Already in 1949, Pauling had shown that sickle cell anemia, a disease in which red blood cells become sickle-shaped, was due to an abnormal structure of hemoglobin, the protein that transfers oxygen through blood to the various tissues (Pauling et al., 1949).[4] In the same year, James Neel (1915–2000) had also shown that sickle cell anemia was inherited as a Mendelian character (Neel, 1949). However, it took a few more years for the connection between the inheritance of a disease and its molecular basis – a specific change in the structure of a protein – to be established. It was eventually in 1956 that Vernon Ingram (1924–2006) showed that the formation of hemoglobin found in patients with sickle cell anemia was due to a mutation in a single gene (Ingram, 1956). This showed how a mutation in a gene could cause a change in a single amino acid, which in turn could have a major impact on the structure of a protein and thus on the emergence of a disease.

Around the same time the conceptualization of genes began to change. The classical genetics conceptualization of genes as beads on a string (Figure 2.3a) implied that genes were indivisible. By the early 1950s the gene had come to be perceived in various distinct ways: as the unit of physiological function, as the unit of mutation, or as the unit of recombination via crossing over. However, it had already been shown that these different kinds of units did not always converge on the same entity (Pontecorvo, 1952). Influenced by this conclusion, and by the then-recent proposal of the structure of DNA by Watson and Crick, Seymour Benzer (1921–2007) (Figure 3.6) thought that the size of genes could be estimated through recombination experiments. He decided to conduct these experiments with bacteriophages and, after a combination of critical events, he made the crucial observation that it was possible to resolve the structure of genes, perhaps even at the level of a single nucleotide. The question that Benzer aimed at answering was whether particular mutations were part of the same functional unit or not. His results showed that the unit of function was larger than both

---

[4] That was the first time that a molecular change was related to a disease, but its impact was not as revolutionary as it has been assumed (see Strasser, 1999, 2015).

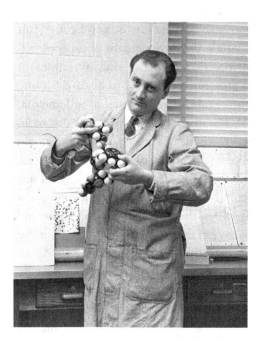

FIGURE 3.6 Seymour Benzer was the first to clearly show the fine structure of genes, and to clarify their different conceptualizations (© Underwood Archives).

the unit of mutation and the unit of recombination. This meant that mutation and recombination *within* a gene were possible. The major conclusion he made was that genes were divisible, and that the main premise for this conclusion was to accept the Watson and Crick model of DNA. The gene thus had a fine structure consisting of a linear array of elements, which were the nucleotides (Benzer, 1955; Judson, 1996, pp. 274–9; Holmes, 2000).

In subsequent work, Benzer (1957) further clarified the new conceptualization of genes: "The classical 'gene,' which served at once as the unit of genetic recombination, of mutation, and of function, is no longer adequate. These units require separate definition" (p. 70). Benzer provided those definitions. He first defined the unit of recombination as the smallest element that was interchangeable but not divisible by genetic recombination. He called this a "recon" and estimated that its size should be no more than two nucleotide pairs. He then defined the unit of mutation as the smallest element that when changed would give rise to a mutant form. He called this a "muton" and estimated that its size should be no more than five nucleotide pairs. Finally, he defined the unit of function as the "cistron," but noted that this was difficult to describe in absolute terms. The cistron was operationally defined based on the results of a specific test, the *cis-trans* test.[5] Its size was estimated to be about 100 times larger than the muton and the recon. Benzer's major contribution was therefore to show how there could be a correspondence between a gene and a sequence of nucleotides on DNA. He concluded: "It should be fascinating to try to translate the 'topography within a cistron into that of a physiologically active structure, such as a polypeptide chain folded to form an enzyme" (Benzer, 1957, p. 93).

Benzer's statement refers to an important process, often ignored today: that a polypeptide chain, i.e. an amino acid sequence, is folded and forms an enzyme. Proteins have a unique way of folding within normally functioning cells. However, in abnormal cells or within microorganisms in a culture dish, different three-dimensional structures are possible for the same molecule. This means that the three-dimensional structure of a protein molecule is not only determined by its amino acid sequence, but also by the external conditions in which the folding of the protein molecule takes place. Therefore, not all the information about protein structure is encoded in the DNA sequence. What is encoded in DNA is the information for the sequence of the amino acids only, which is described as the primary structure of the protein molecule. Its tertiary structure, another term for the three-dimensional structure of a protein, depends on the primary structure as well as on the local conditions. This means that genes do not just make proteins, as it is commonly assumed (see Lewontin, 2000, pp. 73–74). Genes encode the sequences of amino acids of proteins, but not their final three-dimensional structure.

The next big question was how a gene could actually encode the information for a protein. It is not a surprise that when Crick wrote about the nature of the genetic code, he cited Benzer's work as providing the major evidence for it. Crick proposed "The Sequence Hypothesis," which assumed that "the specificity of a piece of nucleic acid is expressed solely by the sequence of its bases, and that this sequence is a (simple)

---

[5] In a *cis-trans* test, the phenotypes of *cis-* (two mutations are on the same chromosome) and *trans-* (two mutations are each on one chromosome) heterozygotes are studied. If mutations *a* and *b* belong to the same cistron, the phenotypes of the *cis-* and *trans-* heterozygotes are different. If, however, the mutations belong to different cistrons, the *cis-* and *trans-* heterozygotes are phenotypically similar (see Holmes, 2000). Under the assumption that most mutations are recessive, from the phenotypes of these individuals one could infer whether or not the mutations were in the same cistron.

code for the amino acid sequence of a particular protein." This served to characterize the relationship between DNA, RNA, and proteins. An important contribution made by Crick in this article was the introduction of the metaphor of information, defined as: "the precise determination of sequence, either of bases in the nucleic acid or of amino acid residues in the protein" (Crick, 1958, p. 153). Thus, DNA, RNA, and proteins were related to one another based on the information encoded within their sequence. These ideas were the foundation for Crick's "Central Dogma": the transfer of information, in the sense of the precise determination of sequence (either of nucleotides in nucleic acids or amino acids in proteins), might be possible from nucleic acid to nucleic acid, or from nucleic acid to protein, but not the other way around. As Crick put it: "once 'information' has passed into protein it *cannot get out again*" (Crick, 1958, pp. 152–153, emphasis in the original).

All these ideas that were proposed by Crick produced a new concept of biological specificity that was based on the transfer of sequence information from one type of macromolecule to another, and thus defined a new intellectual agenda for the emerging discipline of molecular biology (Olby, 1970; Strasser 2006). The previous year, Crick was also the lead author in an article that suggested a structure for the genetic code. Given that there were 4 different nucleotides in DNA (with A, T, C, or G), and that 20 amino acids were found in the proteins of organisms, it was concluded that a single nucleotide could not correspond to a single amino acid, as only 4 amino acids would thus be coded. In addition, two nucleotides for one amino acid would give $4 \times 4 = 16$ combinations, which would still not be enough. Therefore, three nucleotides should correspond to one amino acid. Each of the $4 \times 4 \times 4 = 64$ possible triplets of nucleotides should somehow correspond to each of the 20 amino acids. Crick and his coauthors assumed that the triplets AAA, GGG, CCC, and TTT would make no sense, i.e. would not correspond to any amino acid. This left 60 triplets that, according to them, could be grouped into twenty sets of three as there were 20 amino acids in organisms (Crick et al., 1957).

In 1961, Crick, Sydney Brenner (1927–), who was also well aware of Benzer's work, and two other colleagues drew on experimental evidence to propose several characteristics of the genetic code: that triplets of nucleotides coded one amino acid (they were, however, considering as a less likely event that a multiple of three nucleotides coded one amino acid), that the code was not overlapping (i.e. that consecutive triplets did not share the same nucleotides), that the sequence of bases was read from a fixed starting point, and that the code was probably degenerate (i.e. one amino acid could be coded by more than one triplet) (Crick et al., 1961). All that was fairly accurate, but it was Marshall Nirenberg (1927–2010) and Heinrich Matthaei (1929–) who correctly figured out the correspondence between nucleotides and amino acids. They found that artificial RNAs could stimulate protein synthesis in a cell-free system. The first RNA they tried was poly-U, a long chain of nucleotides containing uracil, and they found that it contained the information for the synthesis of a protein that had many of the characteristics of poly-phenylalanine, i.e. a protein consisting of consecutive molecules of the amino acid phenylalanine. They thus concluded that one U or several Us appeared to be the code for phenylalanine (Nirenberg & Matthaei, 1961). A few more codons were identified over the next year, and by 1965 the genetic code was mostly resolved (see this presented in the form of Table 3.1) (Nirenberg et al., 1965; Söll et al., 1965). The information contained in DNA and RNA sequences can be read as three letter words (triplets) called codons that each corresponds to one amino acid (for the relevant history see Kay 2000; Cobb, 2015). Ironically, as mentioned, the idea that a triplet consisting of the same nucleotide could code for an amino acid had been rejected by Crick and his colleagues (Crick et al., 1957, p. 418). Thus, it was Nirenberg (Figure 3.7) and Matthaei who resolved the genetic code first.

It was around the same time that the chemical nature and structure of genes, and the relation between genes and the genetic code, were understood that the role of mRNA also became clear. This was done in 1961 through the work of François Jacob (1920–2013) and Jacques Monod (1910–1976), who also proposed a model for gene regulation (Figure 3.8). Already in 1955, a connection had been shown between a type of RNA (later called ribosomal RNA or rRNA) and some small structures that were considered as the platform for protein synthesis, called ribosomes (Palade, 1955). Then, it was also concluded that another type of RNA molecule (transfer RNA or tRNA) was a molecule essential for protein synthesis (Hoagland et al., 1958). It was initially thought that the RNA molecules in ribosomes could have a role in protein synthesis. However, later work showed that this was not the case and that another

TABLE 3.1 *The Genetic Code Shows the Correspondence between Three-Letter Combinations (Triplets) of Nucleotides and Amino Acids or Stop Signals (gaps imply that the triplet corresponds to the amino acid right above)*

First Letter	Second Letter U	Second Letter C	Second Letter A	Second Letter G	Third Letter
U	UUU Phenylalanine	UCU Serine	UAU Tyrosine	UGU Cysteine	U
	UUC	UCC	UAC	UGC	C
	UUA	UCA	UAA Stop	UGA Stop	A
	UUG Leucine	UCG	UAG	UGG Tryptophan	G
C	CUU	CCU Proline	CAU Histidine	CGU Arginine	U
	CUC	CCC	CAC	CGC	C
	CUA	CCA	CAA Glutamine	CGA	A
	CUG	CCG	CAG	CGG	G
A	AUU Isoleucine	ACU Threonine	AAU Asparagine	AGU Serine	U
	AUC	ACC	AAC	AGC	C
	AUA	ACA	AAA Lysine	AGA Arginine	A
	AUG Methionine	ACG	AAG	AGG	G
G	GUU Valine	GCU Alanine	GAU Aspartic acid	GGU Glycine	U
	GUC	GCC	GAC	GGC	C
	GUA	GCA	GAA Glutamic acid	GGA	A
	GUG	GCG	GAG	GGG	G

FIGURE 3.7 Marshall Nirenberg was eventually the first to "solve" the puzzle of the genetic code (© Universal History Archive).

FIGURE 3.8 François Jacob (a) and Jacques Monod (b) made the concept of regulatory genes widely considered. (© Henri Bureau).

type of intermediate RNA molecule was involved (Brenner et al., 1961; Gros et al., 1961).

Jacob and Monod also made the important distinction between the "structural genes" that "determine the molecular organization of the proteins" and the "regulator and operator genes" that "control the rate of protein synthesis." According to their model of gene regulation, a particular molecule produced by the regulatory gene could block the expression of the structural genes. Thus, the production of the enzymes encoded in the structural genes did not only depend on those genes themselves, but also on other DNA sequences (Jacob & Monod, 1961). The distinction between "regulatory" genes and "structural" genes was an important contribution of Jacob and Monod. They were perhaps the first to make this distinction more broadly considered, but they were not the first to think along these lines. Barbara McClintock had made suggestions about regulatory (controlling) sequences at least as early as 1953. As soon as the 1961 Jacob-Monod paper was published, McClintock prepared a similar proposal for eukaryotic organisms (McClintock, 1961), which was based on her earlier work (McClintock, 1953, 1956). According to her, the differential expression of genes was the result of their transposition (physical movement) in the genome, which was in turn controlled by other genes. Thus, McClintock had practically anticipated the existence of regulatory genes, but it seems that the key importance that she attributed to the transposition of genes for regulation limited the impact of her conclusions (see also Morange, 1998, pp. 158–159; Keller, 1983, pp. 7–10, 175–178).

Understanding the role and nature of transcription factors was another important advancement of that time (Gilbert & Müller-Hill, 1966; Ptashne, 1967). These are proteins that bind to DNA and facilitate the binding of RNA polymerases so that transcription of DNA to RNA can start. Thus, although all cells of an organism have the same DNA, they have different genes expressed inside them because of the different transcription factors that they contain, which may also depend on the signals they receive from their immediate environment. In particular, transcription factors already exist in the fertilized ovum, mainly derived from the mother, which are then distributed to the several cells emerging from cell divisions. These proteins bind to specific DNA regulatory sequences, called promoters, which are located "before" (upstream of) specific genes, and either activate or suppress the expression

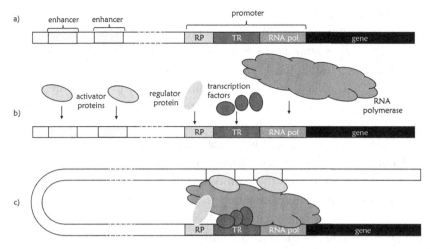

FIGURE 3.9 How transcription starts. Transcription factors and other proteins facilitate the binding of RNA polymerase to the promoter, and then the transcription of the gene can begin (b). This binding depends on DNA sequences both close to the gene and further away (a). The DNA molecule can actually form loops that bring activator proteins bound to enhancers closer to the site of transcription. Then the synthesis of RNA can begin (c). The dotted lines in DNA correspond to long DNA strands that is impossible to depict in this figure.

of these genes. Transcription factors are very specific, i.e. bind to particular DNA sequences, and thus affect certain genes but not others. This also depends on other proteins that bind to other regulatory DNA sequences that may be further away, called enhancers (see Figure 3.9). Transcription factors eventually facilitate the binding of RNA polymerase to the promoter of a gene, and thus an RNA molecule is produced. The transcription of DNA sequences to RNA, which may or may not further be translated into protein (see also Chapter 11), is described as gene expression. When, why, and how gene expression takes place depends on transcription factors and is described as gene regulation. It soon became clear that gene regulation would require more complicated models than the one proposed by Jacob and Monod (e.g. Britten & Davidson, 1969).

We now know that RNA is a molecule quite similar to DNA, but with some differences. The two differences that matter for our discussion is that RNA is a single-stranded molecule and that it contains uracil (U) instead of thymine (T). DNA is transcribed to RNA, and in the case of protein synthesis this RNA is translated into protein. As mentioned, transcription is carried out by enzymes called RNA polymerases, which bind to promoters and produce RNA on the basis of the sequence of the transcribed DNA strand. The RNA produced that carries the information from DNA to the sites of protein synthesis is called messenger RNA (mRNA). This molecule contains a specific sequence of nucleotides that will be translated into a sequence of proteins. This is achieved through a sequence correspondence between mRNA and several tRNA (transfer-RNA) molecules, each of which carries a specific amino acid. These molecules "meet" on ribosomes that consist of rRNA and proteins. For example, a tRNA carries the amino acid methionine and also has the triplet UAC that is complementary to the mRNA triplet AUG (Figure 3.10). Thus, in a sense, it is tRNA molecules themselves that determine the genetic code, as they "link" amino acids and mRNA. However, as shown in Table 3.1, the code is framed in terms of mRNA triplets. Besides the triplets that correspond each to a single amino acid, AUG not only corresponds to methionine but is also the starting point for translation. In contrast, the triplets UAA, UGA, UAG do not correspond to any amino acids, and thus serve as stop signals as translation stops there because no amino acid is added (Alberts et al., 2010, pp. 245–247). This picture of the molecular gene and of its role in the synthesis of proteins was essentially complete by the late 1960s.[6]

The above shows in detail how the "central dogma" that Crick proposed back in 1958 works (Figure 3.11a). However, by 1970 several additions were being made to this model, as it was discovered that the synthesis of DNA from RNA was possible (Baltimore, 1970; Temin & Mizutani, 1970). Crick (1970) commented that these new developments, along with synthesis of RNA from RNA that was already identified in some viruses (Baltimore, 1971), could be easily accommodated by the central dogma. The central dogma could thus describe all kinds of "transfer of sequential information." As Crick noted, the situation

[6] The production of RNA from DNA is described as transcription because during this process the information encoded in the sequence of the nucleotides of DNA (called deoxyribonucleotides) is copied to a similar language, that of nucleotides of RNA (called ribonucleotides). However, the production of proteins from RNA is called translation because the information encoded in the sequence of RNA nucleotides is copied to a quite different language, that of amino acids. The rule of transcription is the specific base pairing of the nucleotides of DNA and RNA (A-T; G-C; A-U), whereas the rule of translation is the genetic code, i.e. the code of correspondence between triplets of nucleotides and single amino acids (Table 3.1), which in turn also depends on base pairing between the nucleotides of mRNA and tRNA (see Figure 3.10).

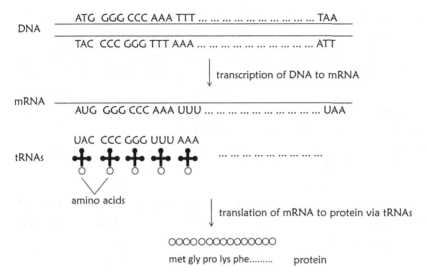

FIGURE 3.10 The flow of genetic information from DNA through RNA to proteins, or more simply the processes of DNA transcription to RNA and of RNA translation to protein (you could check Table 3.1 to confirm the sequence of the protein being produced).

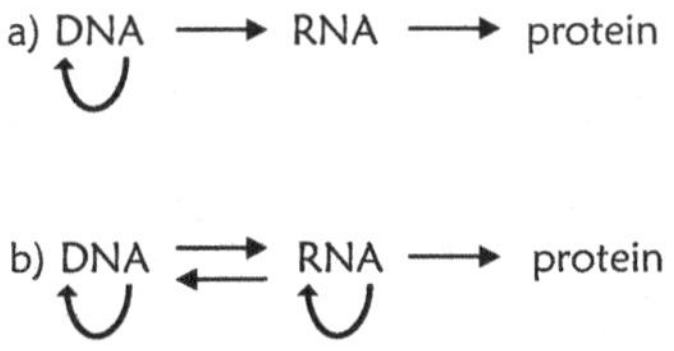

a) DNA ⟶ RNA ⟶ protein

b) DNA ⇄ RNA ⟶ protein

FIGURE 3.11 (a) The central dogma as envisioned by Crick in 1958; (b) The "revised" central dogma as it could be described in the early 1970s. Arrows indicate the "flow" of information on which the synthesis of new molecules is based.

was not yet very clear; however, there was no reason to consider any transfer of information impossible, with the exception of that from protein to other molecules (Figure 3.11b). Thus, by the early 1970s, the central dogma could be seen as summarizing the flow of information from DNA to other molecules, and the gene was thus conceived in molecular terms: as made of DNA, a molecule with a specific structure and particular properties containing sequence information.

On the basis of all these, it is clear that, in contrast to the classical gene, the nature, structure, and localization of the molecular gene was specified, and that therefore the two differed significantly. As explained in Chapter 2, the classical gene was conceived as a factor on chromosomes, whose nature was not specified and was not really necessary to know. It was differences in such factors that caused observed differences in phenotypes, and that was enough for genetic analysis. In contrast, the molecular gene was conceived as a linear sequence of DNA that encoded the information for linear sequences of molecules, either RNA or proteins, which were synthesized through the processes of transcription and translation. Therefore, a main conceptual shift that took place around the middle of the twentieth century was that from the conception of genes as biological determinants without a concrete structure and content to the conception of genes as DNA molecules that carried biological information.[7] However, soon things started to become even more complicated.

---

[7] Lenny Moss (2003, pp. 44–49) has described these two distinct conceptions of gene as Gene-P and Gene-D. Gene-P is defined with respect to a character (e.g. gene for blue eyes, gene for cystic fibrosis), but it tells nothing about the DNA sequence. This is a conceptual tool that mostly refers to what happens when a "normal" allele is missing and has an instrumental utility in predicting a phenotypic outcome. In contrast, Gene-D is defined with respect to a DNA sequence but tells nothing about the observable character. It is a developmental resource with a specific molecular sequence that produces a protein with a specific function. Whereas Gene-P and Gene-D do not correspond exactly to the classical and molecular gene concepts as a molecule can be considered as a Gene-P (Moss, 2003, p. 45), the distinction between the gene conceived as conceptual tool and as a developmental resource is useful for our discussion here (for a similar distinction among gene concepts see also Neumann-Held, 2006).

4   So, What Are Genes?

During the "molecularization" of genes that was described in Chapter 3, the gene concept of classical genetics did not just acquire a material, molecular identity. Neither was it replaced by the molecular concept, or abandoned; in fact, the two concepts coexisted in research, depending on the interests and the explanatory aims of scientists. Nevertheless, the molecular gene concept dominated the scene. It is no coincidence that several of the researchers working on the molecular nature, structure, and expression of genes, such as Watson, Crick, Jacob, Monod, Nirenberg, and Hershey, shared Nobel prizes in Physiology and Medicine between 1962 and 1969. But why didn't the molecular gene simply replace the classical gene? Figure 4.1 provides an illustration of the classical and the molecular gene concepts. As the molecular gene is a DNA sequence that encodes the information for an RNA or a protein molecule thus affecting a character, you might think that there is a clear correspondence between a gene as a discrete DNA segment and a gene as a factor that affects a character, in other words, between the classical and the molecular gene. It would then seem plausible for scientists to replace the former with the latter in research and think of genes as segments of DNA that encode proteins related to certain characters.

However, during the 1970s the complexities of gene expression were revealed in various ways. These made the one-to-one correspondence between genes and gene products (not to mention characters) entirely implausible. The first important complication had already emerged with a better understanding of gene regulation, which was described in Chapter 3. It had been found that changes in DNA sequences that regulate the expression of genes might produce changes in the respective characters. Thus, although the molecular gene that was implicated in the development of a character might remain structurally unchanged, the character could nevertheless change because of changes in the regulation of that gene. In other words, an organism might have a fully functional molecular gene that could produce the protein that in turn would contribute to the development of a corresponding character. But because of a mutation in the regulatory sequence, this gene could be silenced, despite its potential for expression. As a result, it would be possible for

a) classical gene concept (a factor for a character)

b) molecular gene concept (a DNA segment for a protein related to a character)

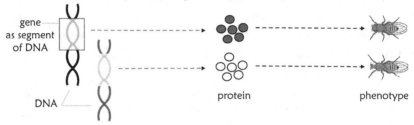

FIGURE 4.1 The classical and the molecular gene concepts may be perceived to have a clear correspondence with each other (between factors and genes, and between phenotypes). However, this correspondence, which is often still taught at schools as a fact, is superficial.

two organisms with exactly the same molecular gene to exhibit different character states because the regulatory sequence of that gene was properly operating in one of them but was somehow switched off in the other.

Let me use an example to illustrate this (see also Figure 4.2). Imagine that there is a switch that can be used to turn on and off the light in a room. In this sense, you can conceive of the switch as controlling whether or not there will be light. Nevertheless, the light is not produced by the switch but by a bulb. Apparently, both the switch and the bulb are necessary in order to have light: the latter "produces" light and the former "controls" its production. If there is both a functional switch and a functional bulb, there is light (Figure 4.2a). If either the switch or the bulb malfunctions, then there is no light (Figures 4.2b and 4.2c). The conceptual problem here is that the classical gene concept could correspond either to the switch or the bulb, whereas the molecular gene concept can only correspond to the bulb. The reason for this is that the nature, structure, and localization of the classical gene concept is not necessary to specify, and so it could be either the DNA segment that

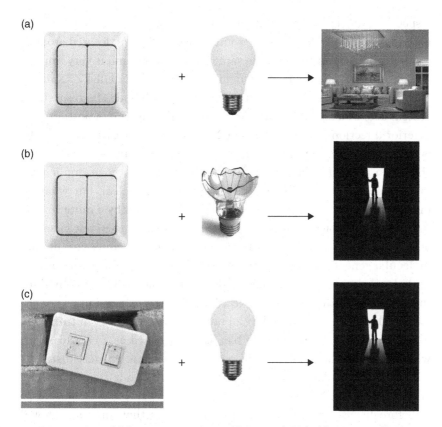

FIGURE 4.2 A model for gene expression and its control. Light (the gene product) requires both a functional switch and a functional bulb in order to be produced. The molecular gene corresponds only to the bulb that produces the light (a). In contrast, the classical gene can correspond either to the bulb (b). or to the switch (c). In this sense, the classical and the molecular gene do not necessarily converge to the same entity. (Photographs © Steven Taylor, Oliver Cleve, Astronaut Images, WIN-Initiative/ Neleman, Iolostock, Achim Sass, clockwise from top left).

produces the respective product (the bulb) or the one that controls its production (the switch). In contrast, the nature, structure, and localization of the molecular gene are necessarily specified, and this could only be the DNA sequence that produces the respective product (the bulb). The conceptual problem, in other words, is that the classical and the molecular gene do not necessarily converge to the same DNA segment. Of course, this is possible, but it is not always the case.

Let us consider an actual example: there is an allele of the gene *Lmbr1/ LMBR1* (limb development membrane protein 1) that produces abnormal limb development in both mice and humans. However, the respective molecular gene that is implicated in limb development is located elsewhere. This gene is "sonic hedgehog" (*Shh*), located around 850,000 nucleotides away on the same chromosome, and is very important in limb development. A sequence within *Lmbr1* called ZRS regulates the expression of *Shh* in the developing limb bud. The respective protein acts as a signal for the patterning of the digits in a posterior to anterior direction (Lettice et al., 2002; see also Hill & Lettice, 2013). It is thus possible for an individual to have abnormal limbs without any change in the *Shh* sequence, which is the molecular gene implicated in limb development. The details are not yet fully clear, but what is important here is that a character may be affected not by a mutation in the respective molecular gene that is directly related to it, but in some other sequence located far away that regulates the expression of the molecular gene (Figure 4.3). As a mutation in *Lmbr1* produces abnormal limb development, it can be considered as the classical gene related to this character. However, the molecular gene, the expression of which is related to abnormal limb development, is *Shh*, which is in turn regulated by *Lmbr1*. This example has been suggested as a clear case in which the classical gene and the molecular gene are entirely distinct; the two genes do not necessarily converge to the same sequences (see Griffiths & Stotz, 2013, p. 61).

Before discussing how the gene concept was further complicated, we should keep in mind that the usual representations of DNA as a double helix (see e.g. Figures 3.2 and 3.4) actually misrepresent the natural state of DNA in cells. DNA is not "naked," but is combined with proteins thus forming the structures that we call chromosomes. These in turn consist of chromatin, a complex molecule that results from the chemical combination – not just the mixture – of DNA and particular proteins called histones. This had become clear already in the early 1970s through the work of Roger Kornberg (1947–) who proposed that chromatin consists of a repeating unit of eight histone molecules and about 200 DNA base pairs (Kornberg, 1974), as well as that two types of histones combined to form a tetramer (Kornberg & Thomas, 1974). The histones are H2A, H2B, H3, and H4, and two histones of each type come together to form what is called a histone octamer (i.e. a structure consisting of eight histone molecules). The DNA molecule is wrapped around this histone octamer two and a half times and the resulting structure is the nucleosome, which is the basic structure of chromatin (Figure 4.4).

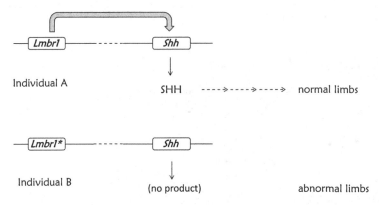

Individual A

SHH ----→----→----→ normal limbs

Individual B

(no product)   abnormal limbs

FIGURE 4.3 It is possible for two individuals to have the same genotype, carrying "normal" alleles of the *Shh* gene, but for one of them to have abnormal limb development due to a mutation in the regulatory sequence that is located within the *Lmbr1* gene. Therefore, individual A has normal limbs because the SHH protein is produced, whereas individual B has abnormal limbs because the allele *Lmbr1** somehow does not activate *Shh* and so the SHH protein is not produced (see Hill & Lettice, 2013). The dotted line between the *Lmbr1* and the *Shh* genes indicates the long distance between them (about 850,000 nucleotides).

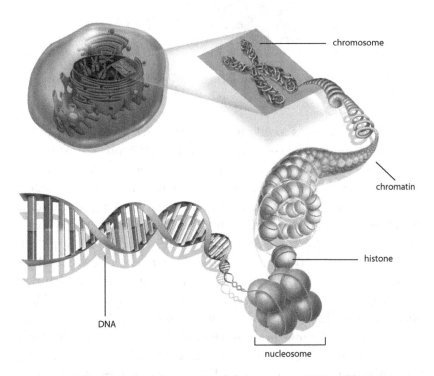

FIGURE 4.4 The structure of chromatin and chromosomes. DNA and histones form chromatin (consisting of consecutive nucleosomes), which condenses before cell division to form chromosomes (© BSIP/UIG).

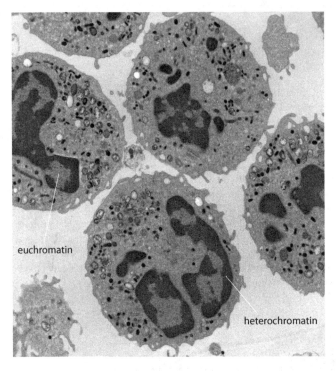

FIGURE 4.5 Chromatin (eu- and hetero-) in the nuclei of a human white blood cell (© Dlumen).

When one observes a cell nucleus in the microscope, it is possible to distinguish between two types of chromatin (Figure 4.5). These are heterochromatin, which is darker and considered as having higher density and less transcriptional activity, and euchromatin, which is less dark and considered as having lower density and higher transcriptional activity. Chromosomes, with the characteristic X-like shape, are observed only during cell division (Griffiths et al., 2005, pp. 80–90; Alberts et al., 2010, pp. 179–188). Therefore, DNA is actually a component of a larger structure, chromatin, with particular biological and physicochemical properties. This structure affects when and how genes are used in multiple ways (see Chapter 11). This implies that DNA is not just "lying" in the nucleus waiting to be "read." It should be noted at this point that Figure 4.4 presents models of the respective molecular and cellular structures. This is not how they actually look, but just how we humans have decided to represent these. But not even Figure 4.5, which is a photograph taken through an electron microscope, shows how things really are, as it is a photograph of cells that have undergone some preparation in order to be observed under the microscope, and not in their natural state. More generally, we should keep in mind the various kinds of interventions that scientists make, and how these affect the status of cells and tissues, before considering anything as "normal" or "natural."

In the 1970s, two puzzling observations were made. The first was that the genome of animals contained large amounts of DNA of unique sequence that should correspond to a number of genes larger than the anticipated one. It was also observed that the RNA molecules in the nuclei of cells were much longer than those found in the cytoplasm. These observations started making sense in 1977, when sequences of mRNA were compared to the corresponding nuclear DNA ones and it was shown that certain sequences that existed in the latter did not exist in the former and must have been somehow removed in the process (Berget et al., 1977; Chow et al., 1977; Sharp, 2005). It was thus concluded that the genes encoding various proteins in eukaryotes included both coding sequences and ones that were not included in the mature mRNA that would reach the ribosomes for translation. These "removed" sequences were called introns, to contrast them with the ones that were expressed in translation, which were called exons. The procedure of removal of intron sequences from the initial mRNA that left only the exon sequences in the mature mRNA was called RNA splicing (Gilbert, 1978; Crick, 1979).

What exactly happens during the process of RNA splicing? Here is a way to illustrate it. Imagine that the following is an mRNA transcript of a DNA sequence that contains the information for the synthesis of a particular structure:

abfhjdkthemvndjklorpeuthaxndjkdoublemvnshasjweuriotnxmclokdfh-snhelixnvmlspaosurjsjshk

At first sight, this sequence seems meaningless. However, if you look carefully you can spot three words that make sense. These are the exons of this gene, whereas the sequences between them are the introns.

abfhjdk**the**mvndjklorpeuthaxndjk**double**mvnshasjweuriotnxmclokdfhsn
helixnvmlspaosurjsjshk

This mRNA contains the specific information for the synthesis of a particular structure. The cell can "extract" this information and produce mature mRNA molecules that will transfer it to the ribosomes of the cell for protein production. These mature mRNAs will include only the sequences that correspond to the exons and will look like this:

the double helix

In this way, the information that is initially "encrypted" in a complex form is transmitted in a clear and straightforward manner to the cytoplasm for the production of proteins. RNA splicing is performed by spliceosomes, ribonucleoprotein complexes that recognize exon-intron boundaries in these precursor mRNA (pre-mRNA) molecules. These remove the introns and assemble the exons together, thus producing the mature mRNA (see also Figure 4.6).

This entails that genes include sequences that correspond to amino acids, but also ones that do not correspond to amino acids. Therefore, a molecular change in a DNA sequence might or might not cause a change in the respective protein. A nucleotide change that took place in an intron might not affect the protein produced, whereas a nucleotide change in an exon might have such an effect. But it would also be possible for a nucleotide change in an exon to produce no change in the character, because e.g. both the old and the new triplet in which the change took place corresponded to the same amino acid. That eukaryotic genes were found to have long noncoding regions that were just removed during gene expression might indicate that only part of a gene, the exons, matter as they contain information. However, this is not the case, as we currently know that introns include sequences that have other roles. One such example has been mentioned previously: the sequence ZRS that affects the expression of *Shh* is located within intron 5 of *Lmbr1*.

What made things even more complicated was the subsequent finding that the same RNA transcript can undergo alternative splicing, and produce different mRNAs that correspond to different proteins produced after translation. In other words, alternative processing patterns were

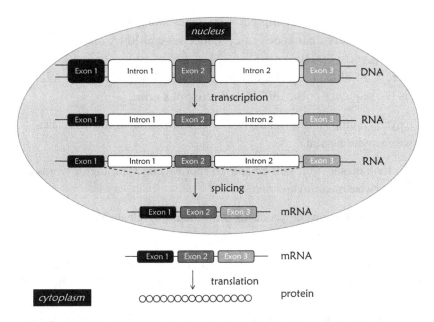

FIGURE 4.6 The process of gene expression in eukaryotic organisms. DNA is transcribed to a pre-mRNA, which is spliced to produce a mature mRNA molecule, which in turn exits the nucleus and is translated to proteins in the cytoplasm (dotted lines indicate the parts of the gene that will be connected after whatever is within the edges of the dotted lines is removed).

possible for a single gene, and these could take place under different cellular conditions, as in the case of the antibodies (immunoglobulins), producing partially similar proteins (Lewin, 1980). Alternative splicing was initially regarded as an exception, but currently more than 90 percent of human genes seem to have at least one alternative splicing event (Pan et al., 2008; Wang et al., 2008). This phenomenon increases significantly the total number of proteins (proteome) produced in organisms (Nilsen & Graveley, 2010). One extreme and impressive case is the *Dscam* (Down syndrome cell adhesion molecule) gene of *Drosophila melanogaster* that can produce 38,016 distinct mRNA molecules (Schmucker et al., 2000)! However, mistakes in this process, or RNA mis-splicing, are also responsible for several human diseases (Scotti & Swanson, 2016). Therefore, the molecular gene is not a linear structure that corresponds to a single RNA or protein molecule, but a modular structure that can be used in different ways to make different molecules.

In this case, the relationship between genes and gene function is one-to-many, as individual genes can produce several RNA molecules and proteins (Lynch, 2004).

Here is an illustration of alternative splicing. Imagine that the following is the sequence of the mRNA transcript of a gene:

nxjyeirhyoufnpwmslonlycndksufjdkslivencmandprldhahahletmxsjlalst-wicencmdoedievnm

If you look closely you will be able to spot several meaningful words in this otherwise meaningless text:

nxjyeirhyoufnpwmslonlycndksufjdkslivencmandprldhahahletmxsjlalst-wicencmdoedievnm

If you are a James Bond fan, you will recognize that the appropriate combinations of these words would produce the titles of two films: the 1967 film *You Only Live Twice* with Sean Connery (1930–) and the 1973 film *Live and Let Die* with Roger Moore (1927–). With plenty of imagination, you could see this gene as the story depository of Ian Fleming (1908–1964), who invented the fictitious MI6 secret agent and wrote the respective stories. This depository could be expressed differentially and produce distinct products (the various stories), which would have some common aspects such as the word "live" in their title (and of course James Bond as the central character). Thus, from the mRNA stemming from this James Bond gene, splicing could produce either mRNA1: **you only live twice** or mRNA2: **live and let die.** It is in this sense that eukaryotic genes contain various elements; these can be combined in different ways and yield different products. Figure 4.7 presents the main types of alternative splicing (see Keren et al., 2010). A concrete example of alternative splicing in humans is the CD44 protein, which is a cell-surface glycoprotein involved in cell-cell interactions, cell adhesion, and migration, and seems to be crucial in cell function. The *CD44* gene has 10 variable exons, i.e. 10 exons that can either all be omitted from the final mRNA or be included in various combinations and produce the respective protein in T cells. Inactivated T cells mainly produce the smallest CD44 protein that lacks all the variable exons. In contrast, T cells activated by some antigen produce multiple CD44 proteins, including those that contain variable exons 1, 3, 4, 5, 7, or 10. In this way, greater specificity of T cells is achieved (Lynch, 2004).

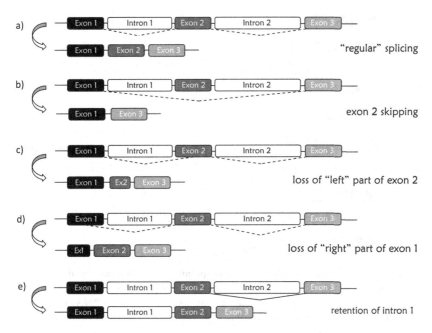

FIGURE 4.7 A eukaryotic mRNA (a), consisting of three exons and two introns. Splicing regularly occurs by removal of the introns, leaving only the exons in the mature mRNA (dotted lines indicate the parts of the gene that will be connected after whatever is within the edges of the dotted lines is removed). There are four types of alternative splicing. In exon skipping (b), an exon is spliced out of the transcript together with its flanking introns; this is the most common type of alternative splicing in complex eukaryotes. In alternative 3′ splice site selection (c) and alternative 5′ splice site selection (d), two or more splice sites are recognized at one end of an exon (3′ indicates the "left" part of the exon and 5′ the "right" part of the exon; "ex" in the figure indicates that only part of the respective exon was retained in the mature mRNA). Finally, in intron retention (d) an intron remains in the mature mRNA transcript. As a result, four different proteins can be produced from exactly the same molecular gene.

Things became even more complicated in the case of trans-splicing,[1] a phenomenon initially observed in unicellular organisms (Sutton & Boothroyd, 1986). During trans-splicing, exons from different mRNAs produced from different genes are joined to form a mature mRNA molecule. In other words, parts of two genes are combined to produce a

---

[1] The "typical" splicing discussed earlier is described as *cis*-splicing. "Cis" indicates that the spliced exons come from the same RNA transcript, whereas "trans" indicates that they come from different RNA transcripts.

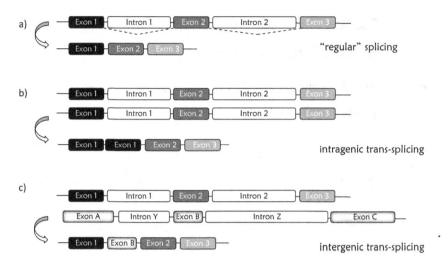

FIGURE 4.8 Intragenic and intergenic trans-splicing. In the first case exons from two pre-mRNAS of the same gene are combined, leading to a duplication of exon I in the resulting mRNA. In the second case, exons from two pre-mRNAs from two different genes are combined, leading to an mRNA that includes exons from both genes.

molecule that might affect a character. Different types of trans-splicing have been described in eukaryotes, including intragenic and intergenic trans-splicing (Figure 4.8). Intragenic trans-splicing occurs when two identical pre-mRNAs from the same gene are spliced together to generate an mRNA with duplicated exons. Intergenic trans-splicing occurs when an mRNA is produced from two mRNAs derived from different genes, from the same or from different chromosomes (Preußer & Bindereif, 2013). Such trans-spliced RNAs are highly expressed in human pluripotent stem cells and differentially expressed during the differentiation of human embryonic stem cells. One of these seems to contribute to the maintenance of pluripotency in human embryonic stem cells by suppressing the expression of genes that are specific in particular cell lineages (Wu et al., 2014).

Even more interesting complications emerged when it was found that the sequences of two molecular genes might overlap. In this case, the same DNA sequence is actually "read" in two different ways and two distinct molecules can be produced. Initially, it was found that in a bacterial virus, mutations for a gene related to the destruction of the bacterial cell that the virus infects were located within the DNA

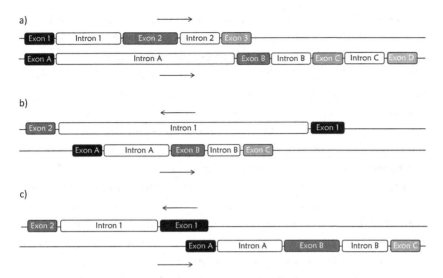

FIGURE 4.9 (a) Overlapping genes on the same DNA strand; (b) Nested genes, where one is included in the intron of the other; (c) Partially overlapping genes over their first exons. In all cases, the arrows indicate the direction of transcription. When the arrows are parallel, the genes are on the same DNA strand; when the arrows are antiparallel, the genes are on the opposite DNA strands.

sequence related to another protein produced in the infected cell. The interpretation given was that the two genes overlapped and that the same DNA sequence was transcribed and translated in two different ways (Barrel et al., 1976). Several overlapping genes have been found in several other organisms, including humans (Veeramachaneni et al., 2004). In many cases, genes are on the same DNA strand, but encode different proteins. In other cases, genes overlap partially over e.g. their first exon sequences (Makalowska et al., 2005). A special case of overlapping genes is the case of nested genes. This refers to a gene that is included within another gene. In eukaryotes, nested genes are usually located within an intron of a host gene (Henikoff et al., 1986). Several nested genes have been found on all human chromosomes. The exons of one gene are usually included within the introns of another gene, and the two overlapping genes are encoded on opposite DNA strands (Karlin et al., 2002; Yu et al., 2005). Figure 4.9 presents some types of overlapping genes. An interesting implication here is that a single molecular change, e.g. a base deletion, might produce two different alleles of the two overlapping genes.

Finally, a phenomenon that further expands this complexity is RNA editing, first observed in the 1980s in protozoa (Benne et al., 1986). It was initially described as "the site-directed modification of the coding potential of genes" (Sommer et al., 1991). More generally, this phenomenon can be defined as any site-specific alteration in the sequence of an RNA molecule, except for changes occurring from processes such as RNA splicing, which results in modifying the sequence of RNA from that of the respective DNA template. RNA editing results in changes in gene expression in all organisms, and can affect the mRNAs, the tRNAs, and the rRNAs in cells. These changes include the insertion of nucleotides, the deletion of nucleotides, and the conversion of one nucleotide to another. In the case of mRNA molecules, such changes can include the creation of start codons through the insertion of bases, changes in the specified amino acids and splicing patterns by base conversion, creation of new sequences by base insertion, and creation or removal of stop codons by base insertion or base conversion. Several changes are possible in tRNAs, including changes in tRNA identity by base conversion at the triplet that "meets" the mRNA, as well as several other changes in the structure of tRNA. Finally, in rRNAs the insertion or conversion of nucleotides can affect translational fidelity and efficiency (Gott & Emeson, 2000; see also Farajollahi & Maas, 2010). A common mechanism of RNA editing in humans is A-to-I RNA editing by ADAR (adenosine deaminases acting on RNA) proteins that convert an A into inosine (I), which is recognized by the cellular machinery as a G. This results in a mature mRNA different from the one encoded in DNA (Nishikura, 2016). This phenomenon is presented in Figure 4.10. Most RNA editing in humans takes place in specific *Alu* sequences (transposable DNA sequences that are called so because of a single recognition site for the restriction enzyme *Alu*I within them), and it is estimated that over 100 million human *Alu* RNA editing sites exist in the majority of human genes (Bazak et al., 2014).

These are only some of the phenomena that pose problems in defining and identifying genes (for these and more see Table 4.1). All these phenomena that made impossible the structural individuation of genes were "discovered" during the 1970s and the 1980s, and before the initiation of the Human Genome Project (HGP) in October 1990. Its primary goal was to identify all genes of the human genome, which were initially estimated to be around 100,000. Additional goals were the sequencing

TABLE 4.1 *Phenomena Complicating the Concept of the Gene*

	Phenomenon	Description	Complication caused
Gene location	Overlapping genes	A DNA sequence may encode two different proteins	No one-to-one correspondence between genes and proteins
	Distant regulatory elements	Sequences located far from the coding sequence affect its expression.	A gene product may be missing without any change in the gene itself.
Gene structural variation	Mobile elements	DNA sequences are found in different locations across generations	DNA sequences do not have a constant location
	Structural gene variants	DNA rearrangements result in alternative gene products	Gene structure may differ across individuals or cells/tissues
	Copy-number gene variants	Variant numbers of DNA sequences	Number of genes may differ across individuals or cells/tissues
Epigenetics and chromosome structure	Epigenetic modifications – imprinting	Inherited information is not DNA-encoded	Phenotype is not strictly determined by genotype
	Effect of chromatin structure	Chromatin structure affects gene expression	Gene expression depends on packing of DNA with protein to form chromosomes

(continued)

TABLE 4.1 *(continued)*

	Phenomenon	Description	Complication caused
Post-transcriptional events	Alternative splicing of RNA	Several mRNAs can be produced by a single transcript	No one-to-one correspondence between genes and proteins
	Trans-splicing of RNA	Parts of the same or different transcript are combined to form an mRNA	No one-to-one correspondence between genes and proteins
	RNA editing	The sequence of RNA is modified	No one-to-one correspondence between genes and proteins
Post-translational events	Protein splicing	The product of translation can be cleaved to produce several protein products	No one-to-one correspondence between genes and proteins
	Protein trans-splicing	Distinct proteins can be spliced together	No one-to-one correspondence between genes and proteins
	Protein modification	Protein structure is modified	No one-to-one correspondence between genes and proteins

Source: Information based on Gerstein et al. (2007, p. 672).

of the approximately three billion nucleotides of the human genome, the development of databases to store this data and of the tools required for their analysis, the consideration of the respective ethical, legal, and social issues, and the sequencing of the genomes of "model organisms," such as the bacterium *Escherichia coli*, the yeast *Saccharomyces cerevisiae*, the roundworm *Caenorhabditis elegans*, the fruit-fly *Drosophila*

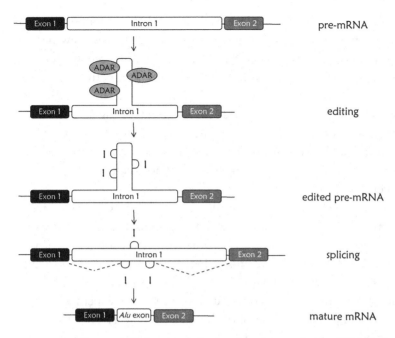

FIGURE 4.10 Inverted Alu repeats in introns from intramolecular RNA duplexes, which are edited by adenosine deaminases acting on RNA (ADARs), changing adenines (A) to inosines (I). The latter are recognized by the splicing machinery as guanines (G), thus creating new splice sites, which results in the inclusion of intronic Alu sequences in the mature mRNAs (this is described as Alu exonization).

melanogaster, and the mouse *Mus musculus* (Gannett, 2014). The HGP was planned to last about fifteen years and complete the sequencing of the human genome around 2005. However, the sequence of the human genome was officially published in 2001 (Lander et al., 2001; Venter et al., 2001).[2] At that point, it was estimated that the human genome contained 30,000–40,000 genes, not 100,000 as it was initially thought.

The publication of the sequence of the human genome was at the end of a tough road, which provides an interesting case of competition in science. The competition was in a sense between the public and the private sector. On the one hand, there were research groups that worked using public funds and that made their data publicly available. These were operating under the auspices of NIH, the Wellcome Trust, and other such institutions, and were led initially by James Watson and

---

[2] If you look at the number of the authors in each of these two articles, you will have a nice illustration of the fact that science does not involve individuals but groups.

after 1993 by Francis Collins (1950–). On the other hand, it was Celera Genomics, founded by Greg Venter (1946–) in 1998, that drew on the already publicly available data to proceed further in sequencing the genome, at a faster pace than the public consortium, without wanting to make their own data publicly available. Venter had already managed to develop new sequencing methods that allowed for the simultaneous sequencing of more than one genes. An intense race thus took place between the public consortium and Celera for two years, until Collins and Venter agreed in 2000 to make a joint announcement (see Davies, 2001; Pennisi, 2001).

As soon as the HGP was completed, it also became clear that the identification of genes on the basis of the human genome sequence alone was not possible, as gene function was also necessary to consider (Snyder & Gerstein, 2003). Then, a new project began: the Encyclopedia of DNA Elements, dubbed as ENCODE. The aim of this project was the generation and the analysis of functional data from various experiments performed on a targeted 1 percent of the human genome. The results of this project made researchers reconsider what a gene is, as the majority of the bases studied were associated with at least one primary transcript, while protein-coding DNA was just 2 percent of the bases studied (ENCODE Project Consortium, 2007). On the basis of these findings, Gerstein et al. (2007) provided the following definition of gene: "The gene is a union of genomic sequences encoding a coherent set of potentially overlapping functional products." This definition summarizes three propositions: (1) that a gene is a genomic sequence (DNA or RNA) directly encoding functional molecules such as RNA or protein; (2) that when there are several functional products sharing overlapping regions, the union of all overlapping genomic sequences coding for them should be considered; and (3) that this union is coherent – i.e., done separately for final protein and RNA products, without the requirement that all products necessarily share a common subsequence (Gerstein et al., 2007, pp. 676–677). This definition is illustrated in Figure 4.11, and according to this a gene is identified on the basis of the functional products that it produces. In other words, it does not matter where the coding sequences are located, but whether their transcripts give combined functional products (either proteins or RNA). Thus, for example, sequences A and C in Figure 4.11 belong to the same gene, although they are further apart from each other compared to the

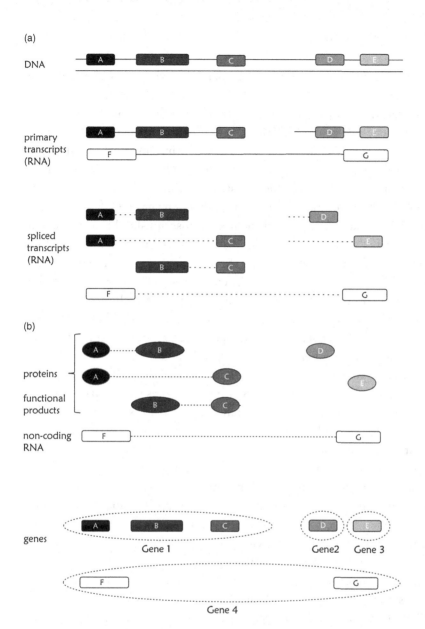

FIGURE 4.11 Identification of genes on the basis of their functional products. A genomic region produces three primary transcripts (ABC, DE, and FG), from which five spliced mRNAs are produced (AB, AC, BC, D, and E). These are eventually translated to produce five corresponding proteins (AB, AC, BC, D, and E), whereas a noncoding RNA also exists (FG). DNA and RNA sequences are depicted as boxes, whereas proteins are depicted as ovals. On the basis of these products, four genes are identified: gene 1, consisting of segments A,B,C; gene 2 consisting of segment D; gene 3 consisting of segment E; and gene 4 consisting of segments FG (based on Gerstein et al., 2007, p. 678; redrawn here with the kind permission of Mark Gerstein).

closely located sequences D and E that are nevertheless considered as different genes.

Given these findings, it is clear that it is implausible to talk about a "one gene-one protein relation," and impossible to structurally individuate genes. Regulatory sequences, discontinuous genes, overlapping genes, alternative splicing, trans-splicing, RNA editing, among other things, have made impossible the structural individuation of genes on DNA. Looking more closely into the phenomena presented in this chapter might make one argue that the RNA transcript should be considered as the "true" gene (Stamatoyannopoulos, 2012). But, this might be like describing the music not based on the score, but on what the musicians play. This in turn would be very difficult to do as the musicians actually rewrite the music the same time that they play it (Keller, 2000, p. 63). The important conclusion from all these phenomena is that DNA does not contain distinct segments corresponding to the genes it is supposed to contain, or, in other words, that genes cannot be structurally individuated. These phenomena can therefore put the existence of genes into doubt (Barnes & Dupré, 2008, p. 55). Do genes really exist? Perhaps they are a heuristic tool for research but nevertheless a human invention that we are still trying to force into existence?

At the same time, it becomes clear that for historical and pragmatic reasons the gene concept cannot just be eliminated. Researchers have been using it and still use it, and they might find hard to let this concept go. Interestingly, in a study of how a small group of biologists conceptualized genes, it was found that whereas a majority of them provided the definition of molecular gene when asked to define a gene, those working in developmental and evolutionary biology also referred to the classical gene concept that focuses on phenotypic characters when asked to describe their actual thinking about genetic problems (Stotz et al., 2004). Therefore, both conceptions can be used in different contexts, and it is sometimes necessary to think of genes in both ways: either as factors that somehow account for differences in organismal characters, or as unobservable entities made of DNA that encode molecules relevant to characters. But no single definition can encompass both concepts. Some have argued that it might make no sense for researchers to shift to other concepts, but to continue using the gene concept although it is hard to define it (e.g. Morange, 2002, pp. 28–32). Others think the opposite: that confusion about genes might be greatly diminished if we talked and wrote about DNA instead (e.g. Keller 2010, p. 77), as well as that in fact there is no molecular biology of the gene but only of the genetic material (e.g. Kitcher, 1982, p. 357).

But even if researchers continue to use this concept, could it perhaps be the case that we should refrain from using the gene concept in science education and communication? For this purpose, it might be useful to replace the concept of "gene" with the broader and more inclusive concept of "genetic material" (Burian & Kampourakis, 2013). This concept can incorporate the various interactions between the genetic material and its intracellular or extracellular contexts, as well as the expression of genetic material to produce RNA, proteins, or other molecules. The concept of genetic material refers to particular macromolecules (DNA, but also RNA), which are related to the expression and inheritance of characters, does not refer to particular functions since it may be implicated in various phenomena and phenotypes, and does not refer to contiguous DNA sequences because functional units may encompass different parts of the genome. The genetic material can be defined as: "any nucleic acid with the propensity to be inherited and to interact with other cellular components as a source of sequence information, eventually affecting or being implicated in cellular processes with local or extended impact" (p. 624). This encompasses all the complexity already presented here, and further in Chapter 11. The public does not need to understand the complexity in all its detail, but only realize that it exists. Therefore, if it is difficult to give a simple and single definition for the gene concept why not use the concept of genetic material that is simpler to define and certainly more inclusive?

There are arguments for both views. Nevertheless, the ENCODE definition of the gene that focuses on the transcript and its functional role seems to be an acceptable one that covers much of the complexity. Therefore, in this book I will consider genes as DNA sequences encoding information for functional products, be it proteins or RNA molecules. With "encoding information," I mean that the DNA sequence is used as the template for the production of an RNA molecule or a protein that performs some function. I generally prefer to refrain from using anthropomorphic metaphors, but "encoding information" has been a widely used expression that makes sense to most people. It is of course not possible to entirely refrain from using metaphors, but we should be explicit about them and appropriately use them in well-defined frameworks, rather than inappropriately building the frameworks on those metaphors.

So far, I have tried to summarize the history of the gene concept from the 1860s until recently. This history can be roughly divided into four periods, corresponding to each of the four chapters so far. The first period from 1864 until the "rediscovery" of Mendel's work in 1900 was characterized by speculative theorizing about hereditary factors. The second period from 1900 until the proposal of the double helix model of DNA was mostly characterized by the classical gene concept, a heuristic tool to do research without necessarily being aware of its nature, structure, and localization. The third period from 1953 until the 1970s was characterized by the molecular gene concept that was conceptualized as a molecular entity capable of carrying genetic information. The fourth period has been characterized by the discovery of a number of phenomena that make a single gene concept untenable. Genes definitely encode some information, but as it will be explained in detail in the next chapters this is not an inherent property. The information that genes may carry makes sense only in an intracellular and intercellular context in which it is expressed. This brings us to the next question of this book. What do genes do?

5   "Genes for" (Almost) Everything

On May 14, 2013, actress Angelina Jolie (1975–) wrote a short essay for the *New York Times*, in which she revealed that she had undergone a double mastectomy for preventive purposes. The reason was that she had been found to carry a "faulty gene," *BRCA1* (breast cancer 1, on chromosome 17),[1] that "sharply" increased her "risk of developing breast cancer and ovarian cancer." In addition, her mother had died at the age of fifty-six after fighting for ten years with cancer.[2] Jolie concluded her essay like this: "I choose not to keep my story private because there are many women who do not know that they might be living under the shadow of cancer. It is my hope that they, too, will be able to get gene tested, and that if they have a high risk they, too, will know that they have strong options. Life comes with many challenges. The ones that should not scare us are the ones we can take on and take control of." Since then, headlines like the following have appeared: "Women like Angelina Jolie who carry the BRCA1 gene are less likely to die from breast cancer if they have their OVARIES removed";[3] "Study: Women with BRCA1 mutations should remove ovaries by 35";[4] "Moms with BRCA breast cancer gene mutations face tough decisions."[5] And there is more, if you do a creative web search.

In order to assess the impact of Jolie's essay, a survey was conducted in the United States with a representative sample of 2,572 adults, who identified their sources of information for this story and also described their understanding of and reaction to it. Participants were

---

[1] Accordingly, there is a *BRCA2* (breast cancer 2) gene on chromosome 13. In particular, *BRCA1* and *BRCA2* alleles have been associated with breast and ovarian cancer. *BRCA1* and *BRCA2* produce proteins that help repair damaged DNA and thus contribute to the stability of the genetic material of a cell acting as tumor suppressor genes. When certain mutations occur, DNA damage may not be repaired properly and so cells are more likely to acquire additional genetic changes that may contribute to the development of cancer.

[2] www.nytimes.com/2013/05/14/opinion/my-medical-choice.html?r=1

[3] www.dailymail.co.uk/health/article-3052396/Women-like-Angelina-Jolie-carry-BRCA1-gene-likely-die-breast-cancer-OVARIES-removed.html

[4] http://edition.cnn.com/2014/02/25/health/brca-study/

[5] www.foxnews.com/health/2015/10/18/moms-with-brca-breast-cancer-gene-mutations-face-tough-decisions.html

asked hypothetical questions regarding preventive surgery in the case of breast cancer, and information about family risk for breast and ovarian cancer was collected. Results indicated that whereas about 74 percent of participants were aware of the Jolie case and about 48 percent could recall Jolie's estimated risk of breast cancer before surgery, fewer than 8 percent had the necessary information in order to accurately estimate her risk of developing breast cancer compared to a woman who did not carry the particular *BRCA1* mutation. It was concluded that participants in this study considered a negative family history as a guarantee against cancer. It was also found that the Angelina Jolie story appeared to confuse participants about the relationship between a positive family history and increased cancer risk (Borzekowski et al., 2013).[6] Unfortunately, estimating and understanding the risks is neither simple, nor straightforward (see Chapter 12).

At the same time, the role of BRCA genes in cancer is highlighted in headlines: "BRCA gene can be a cancer triple whammy, study finds";[7] "Ovarian cancer: we need better access to BRCA testing."[8] Is there a chance that one would adopt a genetic determinist view about disease just because one read headlines such as these? A study aimed at exploring this effect, by randomly assigning ninety-seven college students to three groups who read the same article about research on diabetes. However, students in the first group received the article without a headline, students in the second group received the article with a highly deterministic headline ("Scientists discover gene that causes diabetes"), and students in the third group received the article with a less deterministic headline ("Gene may play a role in diabetes puzzle: germs, genes and diet all contribute to common condition"). The results indicated that all participants had less genetic determinist attitudes after reading the article, with the attitudes among the three groups to be nearly identical. This suggests that the highly deterministic headline did not make those students who received it adopt a similar view. Interviews were also conducted with ninety-five other people (aged between eighteen and forty-five) who read an article with the title "Scientists discover gene that causes diabetes."

---

[6] See Chapters 7 and 12 for more on *BRCA* alleles and cancer genomics.

[7] http://time.com/2810864/brca-gene-can-be-a-cancer-triple-whammy-study-finds/

[8] www.telegraph.co.uk/lifestyle/wellbeing/healthadvice/11565431/Ovarian-cancer-we-need-better-access-to-BRCA-testing.html

After reading the headline alone, participants interpreted the message conveyed as significantly more deterministic than after they had read the full article. These results indicate that there was no framing effect, i.e. that headlines did not frame the interpretation of the article content. However, it is not clear if there is a replacement effect, i.e. if headlines generally stand in for the content of the article because most people do not read the latter (Condit et al., 2001). I am inclined to think that this is quite likely to happen, especially as research in psychology actually suggests that misleading headlines can bias readers toward a specific interpretation (see e.g. Ecker et al., 2014).

Another study aimed at exploring how the public interprets news about behavioral genetics that usually attracts a lot of attention in the news. Overall, 1,413 participants were exposed to one of three published news articles. Of them, 467 participants were given an article on cancer genetics, titled "Key breast cancer gene discovered." These were the control group of the study. The others were given one of two articles on recent findings from behavioral genetics research, titled "'Liberal gene' discovered by scientists," and "Born into debt: Gene linked to credit-card balances," given to 471 and 475 participants, respectively. These formed the two treatment groups of the study. The results indicated that the readers generally accepted the information presented to them; however, they also inadvertently generalized the influence of genetics to other behaviors that were not mentioned in the news article they were exposed to. This means that the members of the public may interpret research on behavioral genetics less cautiously than they do for medical genetics. This may have to do not only with how journalists or scientists present their findings, but also with how nonexperts intuitively interpret what they read (Morin-Chassé, 2014).

But how prevalent are messages that exaggerate the role of genes in the news? A study aimed at assessing public interpretations of popular discourse about genetics, by providing participants with sample genetics news articles and asking for their interpretations of the "blueprint" metaphor. This metaphor refers to a detailed plan used in engineering and architecture and assumes that DNA determines characters by containing such a plan. From the 137 college students who participated in this study, 58 provided responses that were explicitly nondeterministic and 39 provided explicitly deterministic ones (the other participants

TABLE 5.1 *Proportions of Genetic Determinist Statements in Magazine Articles*

Statements in articles	Period			
	1919–1931 (eugenics) (%)	1945–1954 (genetics replacing eugenics) (%)	1967–1976 (molecular biology and genetic counseling) (%)	1985–1995 (medical genetics) (%)
No influence by genes	9	2	0	0
Influence by both genes and environment	61	58	55	73
Influence by genes only	30	40	45	27

Source: Data from Condit et al. (1998); some percentages are rounded to avoid confusion.

provided mixed or irrelevant responses). Interestingly, nondeterministic views were based on interpretations of the "blueprint" metaphor that referred to genes as operating in a partial and probabilistic fashion, as well as being malleable and not determining one's destiny (Condit, 1999). This study therefore suggests that genetic determinist thinking is not as prevalent in the news as one might think it to be. Another study that analyzed the contents of a random sample of 250 American public newspapers and 722 magazines, published between 1919 and 1995, concluded that contemporary descriptions of medical genetics are not more deterministic than those of past times. The study measured three components of genetic determinism in the articles: the degree of genetic determinism (causal influences attributed to genes only, to genes and other factors, or not at all to genes), the type of characters to which genetic causality was attributed (physical, mental, or behavioral), and the degree to which genes were attributed different levels of influence for various characters (strictly genetic or multifactorial). As is evident in Table 5.1, the trend in printed media of more recent times is to include less genetic determinist statements than in the past (Condit et al., 1998).

A more recent study analyzed how the gene concept is presented in major national newspapers from the United States, the United Kingdom, France, and Norway. The study was based on a previously developed framework to distinguish between different ways of framing the gene concept: (1) materialistic, referring to a discrete physical unit; (2) deterministic, referring to a definite causal agent that might even act against environmental factors; (3) relativistic, referring to a predisposing factor; (4) evolutionary, referring to the central object of evolution, a marker for evolutionary change, or a factor that interacts with the environment; and (5) symbolic, referring to an abstract or metaphorical representation of inheritance (Carver et al., 2008). The newspapers were analyzed with respect to the educational level of their readers. Results showed that the symbolic frame was the most frequent one in newspapers with readers of lower education, whereas the evolutionary and the materialistic frame were the most frequent ones in newspapers with readers of higher education. Interestingly for our discussion here, and in line with the conclusions of the Condit et al. (1998) study, there was no overrepresentation of the deterministic frame as it was found in only one-sixth of the genetic discourse. The authors concluded that older accounts of genetic determinism largely concerned symbolic representations of the gene concept, whereas actual claims of genetic determinism were less common in public discourse (Carver et al., 2013).

These studies suggest that overall the impact of genes on characters may not be overemphasized in the press. Could it be the case that the genetic determinist messages are not that many in number, but perhaps that those that exist are strong and influential? Perhaps, but this would require a systematic study of how people interpret the headlines and the news articles they read. This notwithstanding, the situation seems to be quite different in formal science education, where the idea of genetic determinism is quite widespread. For example, a comparative study of biology textbooks from six countries has shown that they present genetics concepts in a way that does not take into account the complexities of development. The model of gene function included in most textbooks was the molecular-informational one, according to which the expression of genes is based on the transfer of information encoded in DNA for the synthesis of RNA and proteins. This model can give a clear molecular explanation of gene function, but is also often explained in a way that reinforces a simplified view of genetic determinism

(Gericke et al., 2014). Interestingly enough, even biology teachers may hold simplistic and inaccurate views of genetic determinism, as a study with teachers from twenty-three countries has shown, with several differences among those countries (Castera & Clement 2014).

Given results like these, it should not be a big surprise to find out that people may finish high school possessing a deterministic view of genetics. This is, for example, the conclusion from a study of 500 high school students' essays, submitted for a national contest, where genetic determinism was found to be a common conception (Mills Shaw et al., 2008). In a study with 383 secondary students, who were asked the question "Why are genes important?", 73 percent responded that genes are involved in the determination of characteristics (Lewis et al., 2000). In another study with sixty-four secondary students, it was found that they map the information included in DNA not only at the level of proteins but also at higher organization levels. In other words, students may map the molecular (genotypic) level directly onto higher levels (which may include that of the phenotype), overlooking the important role of proteins and of all other phenomena and mechanisms that result in the production of characters (Duncan & Reiser, 2007).

These findings suggest that the notion of genetic determinism may not be imposed by the media, but perhaps it is imposed by the simplified representations of gene action in schools. However, there is also evidence that this might be due to more than just imposition. It could be the case that the usual presentation of gene action in schools enhances conceptions that are already held and that stem from deep intuitions.[9] One such intuition is essentialism, the idea that entities have essences, i.e. underlying properties which are characteristic of them (see Wilkins, 2013 for a conceptual analysis). Psychological essentialism is the idea that the essences of organisms are fixed and unchanging, and as such they characterize organisms despite any superficial changes they may undergo (see Gelman, 2003 for a detailed discussion, and Gelman, 2004 for a review). It has been suggested that essentialism has a strong impact

[9] Psychologists write about cognitive biases that give rise to intuitions/preconceptions and eventually to misconceptions. In this book I am using a different terminology, which I find more comprehensible to nonpsychologists: intuitions – a term referring to how we intuitively tend to think – give rise to preconceptions about the natural world, which in turn can change to misconceptions when new knowledge is inappropriately added. Two important intuitions are design teleology and psychological essentialism (see chapter 3 of Kampourakis, 2014).

on people's understanding of genetics. The respective idea has been described as genetic essentialism: "Genetic essentialism reduces the self to a molecular entity, equating human beings, in all their social, historical, and moral complexity, with their genes" (Nelkin & Lindee, 2004, p. 2). In other words, genes are perceived as the essence of being and the source of whatever we are, think, or do, to the extent that it is thus possible to infer one's characters and behaviors from one's genetic makeup. In particular, it has been suggested that genetic essentialism may make people think of genetically influenced traits as unchangeable and determined, consider the relevant genes as being the fundamental cause of the respective character, view groups that share a genetic character as being homogeneous and discrete from other groups, and perceive characters as more natural if they are genetically determined (Dar-Nimrod & Heine, 2011). It has actually been found that people tend to implicitly (and thus unconsciously) associate genes with immutable fate concepts (Gould & Heine, 2012). However, further research on this topic is required.

Another issue is the influence of TV series and films on the public perception of genetics. For instance, it has been found that films may enhance students' preconceptions about genetics. A study with 119 secondary students showed that the concepts conveyed in films about the structure and properties of the genetic material, as well as about its modification, significantly overlapped with students' preconceptions such as their misunderstanding of the nature of genetic material by e.g. not establishing clearly the connection between genes and chromosomes (Muela & Abril, 2014). Films may also relate to ethical issues related to genetics. A well-known film about genetic technologies is Gattaca. This film presents an imaginative future in which, in contrast to the limited current use of human genetic technologies, parents are willing to improve the genetic makeup of their offspring. However, not everyone has access to the respective technologies, and as a result there are two distinct groups of people: the dominant group of the genetically modified and the oppressed group of the genetically unmodified. Interestingly, the film warns that human genetic technologies may cause problems if a genetic determinist ideology that sees humans as the sum of their genes prevails in human societies. However, the film also presents one group of people as genetically inferior to another, and so fails to critically consider problems of basing racial discrimination on differences in genes (Kirby, 2000, 2004).

TV shows seem to have an even more pervasive influence. Even elementary school children consider TV shows as their main source of information about genes and DNA. A study with sixty-two 10–12-year-old children found that they considered television as their main source of information about genetics. The majority of these children knew about DNA (89 percent) and genes (60 percent); but they also thought that DNA is found only in blood and body parts used for forensic investigations (Donovan & Venville, 2014). Such findings are especially important, given that popular TV series such as "Crime Scene Investigation" (CSI) seem to advance an entirely unrealistic view of what can be achieved in science laboratories. As a result, they impose a demand and a respect for forensic science that is not supported by its actual validity (Schweitzer & Saks, 2007). A detailed study of this TV series concluded that its predominant message is that DNA testing is common, swift, reliable, and instrumental in solving cases. It was also found that the show presents DNA processing as quick and easy, typically taking no more than a day to complete, whereas this process may actually take days, weeks, or months (Ley et al., 2012). Consequently, if TV shows generally portray the conclusions of genetics testing as definitive and certain, and this view is easily accommodated by the public, the latter will come to have unrealistic expectations and a wrong perception of the impact of genetics knowledge on societal issues (see also Chapter 12). This is crucial, especially as research suggests that TV shows that impose a view of science as certain enhance people's interest in it, whereas those that impose a view of science as uncertain may not have such an effect (Retzbach et al., 2013).

Several other studies have attempted to document the public's understanding of and attitudes toward genetics. A review of studies on adults' understandings of the role of genes in human health has shown that their understanding has some overlaps with that of professionals, but that it also has many inconsistencies. The majority of people studied seem to understand genetics in terms of heredity and the transmission of characters across generations in families, rather than in terms of gene structure and function. Their understanding of heredity, in turn, is influenced by their understanding of social relationships, as well as by their familiarity with particular conditions through experience. Several studies have also concluded that people tend to believe that both genes and other factors influence many health conditions.

Nevertheless, people do not understand gene-environment interactions well, and this is in part due to difficulties in understanding probabilities. They may also think of disease, and disease prevention or cure, in terms of their common sense understandings of how body systems work, linking e.g. smoking strongly to lung cancer because their relation is obvious, but less strongly to heart disease because their relation is less obvious. Finally, several studies have shown that many people have a vague understanding of genetic testing by confusing it with nongenetic tests, such as e.g. those for cholesterol, and by anticipating that a genetic test might indicate whether or not one is at risk for a condition. Interestingly, it seems that this vague understanding is associated with extremely highly favorable views of most genetic tests (Condit 2010a).

Another review focused on studies of attitudes toward genetics. It aimed at providing a synthetic perspective of the relevant literature on favorable or unfavorable attitudes toward genetic technologies and their applications. It was concluded that attitudes toward genetics are influenced by perceived personal benefits and harms, thus resulting in apparently contradictory beliefs about genetics. For instance, people might suggest that genetic tests should be made available to those who want them, and at the same time that such tests might lead to some kind of discrimination. These apparent contradictions seem to reflect different combinations of costs and benefits made in different contexts. Attitudes toward genetic testing and pharmacogenomics were generally positive, those toward genetically modified food were ambivalent, and those toward cloning were negative. These attitudes varied by education level, and to some extent by gender and ethnicity, but not in a simple fashion (Condit, 2010b).

An important question here is what kind of correlation exists between a good understanding of genetics and the respective attitudes. A study with 560 women who had been offered prenatal screening measured knowledge about, and attitudes toward, genetic testing and the uses of genetic information. Overall, participants supported genetic testing for improving disease diagnosis. There was a small, but statistically significant, correlation between knowledge and attitude, such that respondents with higher knowledge scores had a more positive attitude toward genetics. However, participants with a better knowledge of genetics also held more critical attitudes toward morally contentious issues (Etchegary, 2010). Therefore, a better-informed public is not necessarily supportive of all genetics applications and may have a critical attitude toward them. This would be very good if it were generally the case, because it would allow for a better evaluation of the costs and benefits of genetic testing. This evaluation is very important, yet not simple and straightforward.

With people finding the idea that there are "genes for" characters very intuitive, and with TV series giving the impression that genetic tests are easy to do and that they provide definite conclusions, there exist companies that sell direct-to-consumer genetic (DTCG) tests that make a variety of promises – the most important one being that they will guide you to live a better and healthier life. On what are such conclusions based? There exist numerous studies of associations between specific genes and disease. In such studies, researchers study two distinct groups, one including people having the same disease (the "disease" group) and one including healthy people who do not have the disease (the "control" group). Then researchers look for particular alleles of particular genes, and try to see whether these are more common in one or the other group. If an allele is more common in the "disease" group, it can be said to be associated with the disease; if an allele is more common in the control group it can be said to be protective for the disease (these studies are discussed in detail in Chapter 6).[10] Companies selling DTCG tests draw on the conclusions of this kind of research to explain the results of tests of individual patients.[11] Let us now see in more detail what these DTCG-test companies claim to offer.

Several companies of this kind were launched between 2007 and 2009, including deCODEme, Navigenics, 23andMe, and Pathway Genomics (Davies, 2010). The first two no longer operate, whereas the latter two still do.[12] 23andMe promises a better understanding of ourselves: "Get to know you. Health and ancestry start here. View reports on over 100 health conditions and traits; Find out about your inherited risk factors

[10] The distinction between association and causality in genetics research is discussed in detail in Chapter 8.
[11] Some of these companies have health professionals to explain the results, whereas others do not. Some of these companies accept orders only from physicians, whereas others accept orders directly from consumers. These are important differences, but I do not consider them here in order to refrain from complicating the discussion.
[12] All excerpts from the websites of the companies selling genetic tests were taken on September 1, 2016. It is possible that the respective texts have changed after that date.

and how you might respond to certain medications; Discover your lineage and find DNA relatives" (www.23andme.com/en-eu/). When it comes to health, quite careful statements are made: "Our genes are a part of who we are, so naturally they impact our health. By knowing more about your DNA, you may be able to take steps towards living a healthier life. Keep in mind that many conditions and traits are influenced by multiple factors. Our reports are intended for informational purposes only and do not diagnose disease or illness" (www.23andme .com/en-eu/health/). On that same health page, it is stated that "23and-Me's in-home, saliva-based service helps you know more about yourself. With reports on over 100 health conditions and traits, here are a few of the things you can learn about you." These include forty-plus reports for inherited conditions, ten-plus reports for drug response, ten-plus reports for genetic risk factors, and forty-plus reports for characters, and there is a page for each of these. In all these pages, one finds the same reminder: "Our reports are intended for informational purposes only and do not diagnose disease or illness." Overall, 23andMe does not claim to predict anything about health and disease.[13]

Another company, Pathway Genomics (www.pathway.com), focuses on human diseases. On the webpage of its Cardiac DNA Insight® test, it states that this test "analyzes patients' DNA to identify specific genetic variants associated with an increased risk of developing certain heart-related health conditions such as hypertension, atrial fibrillation and myocardial infarction. This test also provides insight into patients' potential responses to eight classes of commonly prescribed medications that are used to treat heart-related conditions or are known to affect the cardiovascular system" (www.pathway.com/cardiac-dna-insight/). On the page of another test, BRCATrue®, we read that it "searches for mutations in BRCA1 and BRCA2 genes. Having mutations in either the BRCA1 or the BRCA2 gene significantly increases a patient's risk for breast, ovarian and other types of cancer" (www.pathway.com/ brcatrue/). Finally, another test, Mental Health DNA Insight® "analyzes patients' DNA to identify specific genetic variants that can affect how they respond to over 50 psychiatric medications ... indicated for major depressive disorder (MDD), schizophrenia, bipolar disorder, epilepsy,

---

[13] This makes more sense if one takes into account an earlier FDA (Food and Drug Administration) decision about a genetic test marketed by 23andMe (see Chapter 8).

seizures, attention deficit/hyperactivity disorder (ADHD), anxiety and other neurological disorders" (www.pathway.com/mental-health-dna-insight/). The statements on the website of this company are carefully articulated and do not make unreasonable promises.

A UK-based company called GeneticHealth (www.genetic-health .co.uk) initially seems to make equally careful promises:

> Medical research is increasingly clarifying the link between an individual's genetic makeup and their risk of many age related diseases. The interaction between your genes and the way you lead your life, "your environment," can play a crucial role in determining your future health. At GeneticHealth, we provide state of the art genetic testing that specifically looks at a group of selected genes that have been shown by medical research to be implicated with your susceptibility to several age related diseases. We look at this genetic data in context with your family history and lifestyle and advise you on your level of risk for each of the major disease areas.

So, genes interact with life habits, and are implicated to the development of disease. What genetic tests can do is advise one on the possible risk to develop a disease; nothing more. However, on another page we read (www.genetic-health.co.uk/dna-test-services.htm): "The genetic testing studies at a number of genes and more specifically examines variations called *polymorphisms*. These variations exist in all of us. As well as determining our unique features such as eye and hair color, they also determine how our body functions internally. A gene profile, for example, can predict your predisposition to heart disease, determine whether you have a sensitivity to putting on weight or susceptibility to different cancers." Note the verbs used in this statement: variations "determine" unique features and functions of the body, and the gene profile can "predict" predispositions and susceptibilities. "Determine" means settle, decide, or cause as if there is no other influence; "predict" means see in advance with some level of certainty. The authors of this statement might have used other verbs, such as "affect" instead of "determine" and "estimate" instead of "predict," which would indicate less certainty. Thus, the message conveyed in this case is different from those of the companies mentioned above.

Myriad Genetics Inc. (www.myriad.com) claims to *assess* risks, rather than predict them: "The BRAC*Analysis* ® test assesses a person's risk of developing hereditary breast or ovarian cancer based on the detection of mutations in the *BRCA1* and *BRCA2* genes"; "COLARIS® testing assesses a person's risk of developing hereditary colorectal cancer and a woman's risk of developing hereditary uterine/endometrial cancer by detecting disease-causing mutations in the *MLH1*, *MSH2*, *MSH6* and *PMS2* genes"; "COLARIS *AP* ® testing assesses a person's risk of developing hereditary colorectal polyps and cancer by detecting mutations in the *APC* and *MYH* genes"; "MELARIS® testing assesses a person's risk of developing hereditary melanoma by detecting inherited mutations in the *p16* gene (also called *CDKN2A* or *INK4A*)"; "PANEXIA® testing assesses a person's risk of developing hereditary pancreatic cancer by analyzing the *PALB2* and *BRCA2* genes, which are most commonly mutated in families with hereditary pancreatic cancer." And there is more (www.myriad.com/products-services/all-products/overview/). To estimate this risk, the company provides a quiz (www.hereditarycancerquiz.com). What is not discussed in detail here and in the previous websites is how exactly the risk should be estimated and understood (see Chapter 12 for a concrete example).

Other companies help you determine your ancestry (23andMe seems to have recently focused on this, too). One of these companies is iGENEA, which provides a test that makes possible the determination of "your ancient tribe and your region of origin" (www.igenea.com/en/home). This test helps you identify where you come from and defines "ancient tribes" as "peoples from ancient times who are defined not only by their own language, culture and history, but also by their own DNA profile" (www.igenea.com/en/origins-analysis). In this way, the company states, one can identify one's deep origins, which may matter if e.g. one is not aware of the details of one's genetic origins because e.g. of having been adopted. In this case "A DNA test confirms or refutes your assumptions and provides you with new clues and links with relatives about whom you previously knew nothing" (www.igenea.com/en/adopted). Furthermore, a more diachronic service is offered: "In addition to a certificate containing an origins analysis, genetic profile and personal interpretation, you also receive permanent and unlimited access to the largest DNA genealogy database in the world, which enables you to find people who share common ancestors with you" (www.igenea.com/en/home). In other words, the results of this test provide people with information that will allow the precise determination of one's ancestry, by providing information about distant ancestors. This is a different kind of interpretation of genetic data that looks to the past and what has happened, not to the future and what might happen as it is the case with tests about risk to develop a disease. In this case, one can make conclusions about past events rather than trying to predict the future e.g. whether or not one might develop a disease.

Finally, certain companies offer some really impressive services. The company GenePartner has developed the GenePartner formula that "measures the genetic compatibility between two individuals and makes an accurate prediction of the strength of their basis for a long-lasting and fulfilling romantic relationship." The measurement is based in part on genes that encode some proteins of the immune system, called human leucocyte antigens (HLA), found on the surface of human white blood cells, and involved in the cell recognition processes of the immune system. According to the company, "Genetic compatibility results in an increased likelihood of forming an enduring and successful relationship. Research has also shown that the sex lives of genetically compatible partners are more satisfying than average. Additionally, fertility rates are higher in genetically compatible couples and they have healthier children" (www.genepartner.com/index.php/science). In short, the company claims to measure the genetic compatibility of two people, and to accurately predict how strong and successful their relationship will be. Perhaps the first date should be in their offices and not for a coffee to get to know each other? Should all married couples take this test in order to see if they are really compatible? Perhaps the problems that certain couples face are not due to their behavior but due to their genetic incompatibility?

As it is obvious from the examples shown, the companies selling genetic tests convey variable messages regarding the potential of these tests. An analysis of twenty-nine health-related direct-to-consumer websites aimed at examining whether their informational content (e.g. benefits, limitations), literacy demands (e.g. reading level), and usability (e.g. ease of navigation) met existing recommendations. The main conclusion was that most sites provided information about the health conditions and the markers for which they tested, the benefits of testing, the testing process, and their privacy policy. Most sites listed at least one benefit for consumers undergoing testing, the most common one being that the test results can inform a health decision. However, only eleven websites provided any scientific evidence in support of their tests and, further, only six of those cited the relevant scientific literature. Regarding limitations, about half of the websites presented at least one limitation related to testing, the most common being that other factors are important for disease risk besides genetic variants. Most interestingly, only about half of the websites mentioned that the science underlying the tests is new and/or changing, and this was not usually described as a limitation of the tests. It was overall concluded that many users would struggle to find and understand the important information, understand the content on these sites and appropriately interpret the results of the respective tests (Lachance et al., 2010).

Research also suggests that the public does not seem to be appropriately informed about the applications of these tests and the conclusions that can be drawn from them. A review of seventeen studies on users' perspectives on DTCG tests has revealed a low level of awareness, as well as concerns about the reliability of these tests and the respective privacy issues. It should be noted, though, that only two of these seventeen studies reviewed involved actual users of such tests. Participants in these studies raised two main concerns: privacy issues and the nature of the results and their impact (Goldsmith et al., 2012). A recent study that compared how 145 individuals from the general public and 171 genetic counselors interpreted results from DTCG tests suggests that there are important differences between the two groups, and that the general public may misinterpret the results of these tests. The findings of this study suggest that at least some members of the public will need guidance to understand the conclusions that can be drawn from their test results (Leighton et al., 2012). Therefore, as these genetic tests are widely available, the public should be appropriately informed about their actual applications, the possible conclusions, and the relevant ethical issues (see Kampourakis et al., 2014 for an overview; a very informative resource is Annas & Elias, 2015).

One might then anticipate that at least physicians and other health-care professionals could effectively advise the public about these tests. However, this does not seem to be the case either. For instance, a national US survey of 10,303 physicians, a sample considered representative of the overall US physician population, indicated that only 29.0 percent of them had received any education in pharmacogenomics and only 10.3 percent felt adequately informed about pharmacogenomic testing. What these physicians seemed to lack was knowledge about what tests are available, when to use them, where to obtain them, how to interpret the results, and how to apply them in individual cases (Stanek et al., 2012). Another study involved 597 primary care physicians in the United States. The assumption behind this study was that as the number of drugs for which testing becomes available increases, more and more primary care physicians will have to use such tests, even if they do not order it themselves but through a clinical specialist. It was found that only 13 percent indicated that they felt comfortable ordering such tests, and 16 percent reported that they had received some formal education about pharmacogenomics. In addition, 20 percent of them had ordered a relevant test at least once a year in the past (Haga et al., 2012). Finally, and perhaps most interestingly, a study with 516 clinical geneticists and genetic counselors indicated that even these people might not be fully prepared for pharmacogenomics counseling. Twelve percent of genetic counselors and 41 percent of clinical geneticists indicated that they had ordered or coordinated patient care for pharmacogenomics testing. However, and despite the fact that almost all respondents had some education on pharmacogenomics, only 28 percent of counselors and 58 percent of clinical geneticists indicated that they felt well-informed (Haga et al., 2012).

Therefore, it seems that there is a need to better educate healthcare professionals. Recommendations and ideas for this purpose are already available, such as: (1) increasing the amount of content that is related to genetics and common diseases as opposed to rare Mendelian diseases in the preservice curriculum; (2) ensuring that instruction is case-based and reflects practical examples that demonstrate that genetics matters on a daily basis and can improve patient outcomes; and (3) developing continuing education programs (Guttmacher et al., 2007). All of these recommendations are important. First, the fundamental knowledge of genetics that healthcare professionals receive should not be limited to typical Mendelian genetics, but should carefully consider current genomics research. Second, the cases discussed should reflect the actual practice; rare diseases are important for those who carry them, but being able to explain the actual risk to a person with a family history of cancer and a *BRCA1* allele will perhaps be useful to more people. Finally, as genomics research is developing and changing at a fast pace, continuing genetics education programs for physicians and other healthcare professionals are necessary for them to be able to follow the current developments.

There is an additional recommendation that I would add to this list: clarifying crucial concepts and carefully explaining both to the public and to the physicians and healthcare professionals exactly what the state of the art is at each given time. This in turn depends on the terms and the concepts used. In the case of recent genomics research the terms that have been used may have – in part – caused the misunderstanding of the actual potential of genetic testing and analyses described in this chapter. Therefore, one crucial point to clarify is that DNA sequencing is entirely different from DNA decoding, although it is a prerequisite for it. To understand this, it is useful to make clear what kinds of technologies are currently available for "reading" DNA sequences. In the procedure called whole genome sequencing, the sequences of short DNA fragments are determined and are then compared to one another. On the basis of their overlapping parts, the sequence of these fragments is estimated. The end-result is a whole genome sequence that is then compared to a previously developed reference genome, which is nevertheless incomplete as some regions are hard to sequence. It must be noted that errors are possible in these procedures, and so each base is usually sequenced about thirty times (and a lot more when the aim is to identify a disease-associated variant). An alternative approach is called exome sequencing, and in this case the aim is to find the sequence of the exons of genes. Apparently, this is less costly than whole genome sequencing, but it may also miss important information about variants found outside protein-coding genes (Snyder, 2016, pp. 13–18). In either case, it is possible to find the sequence of DNA (in terms of the nucleotide bases ATCG) without necessarily knowing if variations at certain parts are important or not.

In other words, whole genome and exome sequencing are just the means to produce a sequence of bases, a "text," which correspond to the genome or the exome of an individual. Once such sequencing is completed, we can certainly try to read the "text" produced; however, this does not mean that we can always make sense of it. This, in turn, requires the ability to decode the text, i.e. find words and meaning in a long sequence of letters, as the example that follows (already discussed in Chapter 4) illustrates:

Outcome of sequencing:

abfhjdkthemvndjklorpeuthaxndjkdoublemvnshasjweuriotnxmclokdfhsn-helixnvmlspaosurjsjshk

Outcome of decoding:

abfhjdkthemvndjklorpeuthaxndjkdoublemvnshasjweuriotnxmclokdfhsn-helixnvmlspaosurjsjshk

In this example, the outcome of sequencing is a long sequence of letters that at first sight make no sense, in other words, does not seem to encode any message or information. Making sense of this sequence requires an additional procedure after sequencing, which is about finding which DNA segments are genes or regulatory sequences. This is the decoding of the genome and it is a lot more difficult to do than simply finding the sequence, which is not simple anyway.

We are still in the course of decoding the human genome, fifteen whole years after the publication of its sequence in 2001, and still far from achieving it. Important progress has of course been made, but there is still a long road ahead. However, as shown in Figure 5.1, some people were already optimistic in 2000, when the completion of the sequence of the draft human genome was officially announced. The statement on the screen in the background should have been "sequencing the book of life," or even better "sequencing the human genome," and not "decoding the book of life." Decoding the human genome was then the next step, and it seems that even today there is still a lot that researchers need to do in order to achieve it. In other words, the message conveyed in the announcement of what the screen in Figure 5.1 also describes as "a milestone for humanity" did not accurately reflect the actual achievement. Was it unconscious? Was it hype? Whatever the reason, the problem is that with messages like this, it is natural for nonexperts to have great expectations, and for commercial enterprises to try to meet these expectations by selling people what they would like to have. And this was an official announcement, in which the leading scientists were involved; no journalists or others are to blame. President Clinton famously proclaimed: "Today we are learning the language in

FIGURE 5.1 The announcement of the sequence of the human genome on June 26, 2000, was characterized by wishful thinking about what we would like to achieve ("decoding the book of life," as shown in the photo), rather than what was actually achieved at the time. From left to right, Craig Venter, who represented Celera Genomics, Bill Clinton, who was then the president of the United States, and Francis Collins, who represented the public consortium to sequence the human genome (© Joyce Naltchayan).

which God created life."[14] I am afraid not. On that day, we only learned the sequence of letters but we still need to figure out the details of the syntax and the grammar. This is why scientists should be very cautious in how they communicate their findings to the public (a point to which I return in my Concluding Remarks).

From the research presented in this chapter, it becomes clear that the public is not as informed as it should be about the conclusions from genetics research and the respective applications such as genetic testing. The idea of genetic determinism, even if it is not prevalent in the media, certainly is prevalent in formal education and seems to be quite intuitive. This, combined with the fact that the public receives from the media a rather distorted portrayal of what DNA analysis can

---

[14] https://partners.nytimes.com/library/national/science/062700sci-genome-text.html

reveal, may give rise to high expectations from genetic testing – or at least higher than what is currently possible to achieve. This is especially important as certain enterprises selling genetic tests do not always explain what the genetic tests can actually provide and their limitations, and as healthcare professionals do not seem to be prepared to help in this direction. Now, in order to make sense of the results of such tests, we need to have a clear idea of how genes relate to characters and diseases. For this purpose, in the next two chapters I explore whether or not the idea of "genes for" characters and diseases accurately represents biological phenomena. Let us then see what genetics research has so far revealed for the genetics of human characters and diseases.

6   Are There "Genes for" Characters?

It is now time to see whether the idea of "genes for" is correct, i.e. whether it accurately represents biological phenomena. Inherited characters are generally conceived as being of two major types: (1) simple ones, which are considered as determined by single genes, and (2) complex ones, which are considered as being affected either by several genes or by one or more genes and environmental factors. In the first case, reference is usually made to monogenic characters and it is assumed that certain alleles determine one or the other version of the character under discussion. Typical textbook examples are the shape and color of Mendel's peas, or the eye color in humans. In the second case, it is assumed that several genes are involved in the production of a character, with environmental factors also being involved in many cases. Such complex characters can be polygenic, as in the case of human height. In other cases, such as human aggression, it is assumed that both genes and environment have an influence, implicitly giving priority to genes in many cases. I discuss each of these characters as examples in some detail in this chapter, in an attempt to clarify what the actual impact of genes is upon them.

Let us start with monogenic characters. Typical examples of this kind are the characters of peas that Mendel studied, such as the form of the seed (round/wrinkled) and the color of the seed (yellow/green) that were presented in Chapter 1. However, although textbooks go as far as to imply that Mendel knew which alleles were related to each of these characters, this has only recently started to become clear. Wrinkled seeds have higher amounts of sucrose, fructose, and glucose, resulting in higher water uptake due to osmosis (the phenomenon during which there is more water transfer from lower to higher concentrations than in the opposite direction). It has been shown that a form of the enzyme SBE1 (starch-branching enzyme), which is involved in starch synthesis, is missing in wrinkled seeds. The *r* allele results from the interruption of the *SBE1* gene by an insertion of 800 nucleotide pairs. Because of the lack of the SBE1 enzyme, there is reduced starch biosynthesis and so higher amounts of sucrose, fructose, and glucose (Bhattacharyya et al., 1990). Seed color (yellow/green) is regulated by alleles *I* and *i* (the notation Y and *y* was used in Chapter 1 for clarity, as Y more clearly indicates yellow color), with the latter retaining seed greenness not only during seed maturation but also during senescence (when cells lose the ability to divide). This suggests that allele *i* is related to the *stay-green* (*SGR*) gene, which encodes the SGR protein that is involved in the chlorophyll catabolic pathway (the exact function is not clear). Three different *i* alleles have been found; one results in the insertion of two amino acids in the SGR protein, whereas the other two result in low or no production of this protein (Sato et al., 2007). Overall, it has been shown that the genes related to the characters that Mendel studied are not entirely independent as he assumed, and that the alleles result from a range of changes at the molecular level (see Reid & Ross, 2011, for an overview).

An important first conclusion is that the genes related to Mendel's characters are not "genes for" these characters. There is no "gene for" wrinkled peas; the shape of seeds is the outcome of the function of an enzyme that synthesizes starch; wrinkled peas are produced because of physicochemical phenomena (osmosis) and not because there is an allele that determines this shape. Similarly, the color of seeds depends on the metabolism of chlorophyll and the proteins involved in that. Green color is therefore not produced by an allele "for" this color, but when the regular chlorophyll metabolism changes because of changes in the respective proteins. In other words, these two characters that Mendel studied are not determined by specific genes, but are affected by genes producing proteins involved in plant metabolism. It is changes in these genes that produce changes in the respective character, as Morgan and his colleagues had concluded for various characters in *Drosophila* (see Chapter 2). In this sense, wrinkled shape and green color are the by-products of altered metabolic reactions, and not the products of any "genes for" seed shape or color. Therefore, genes certainly affect characters but do not directly determine them.

Before proceeding further, let us see how the different alleles come to existence. The phenomenon through which new alleles are produced is mutation, which results in changes in the sequence of nucleotides. It must be noted that mutation means "change", nothing more; whether this change has a good or a bad outcome is a different story. In the simplest case, there can be a change in a single nucleotide. This change can be the result of a mistake during DNA replication (see Chapter 3). Such a mistake may not affect the message at all; it may affect it a little; or it may change it significantly. Consider the title of a famous Beatles' song: "I Want To Hold Your Hand." A change in this sentence is possible that would replace the period with an exclamation mark: "I Want To Hold Your Hand." This change does not really affect the message, although it changes the emphasis. However, you would still express your feelings to your beloved person; that would be clear. Now, a change in a single letter might have a different outcome: consider the replacement of "H" by "F" in "Hold." The message would then become "I Want To Fold Your Hand," and your beloved person would have good reasons to be concerned about your intentions. It is in such ways that single nucleotide polymorphisms (SNPs, often pronounced as "snips"; they are also described as SNVs, i.e. single nucleotide variants), i.e. particular sequences varying in a single nucleotide, also occur. For instance, the sequence ...CTA<u>A</u>GTA... might change to ...CTA<u>G</u>GTA... SNPs are distinguished from rare variations by the requirement for the least abundant allele to have a frequency of 1 percent or more (Brookes, 1999).

Except for the characters that Mendel studied, another typical textbook example is the inheritance of human eye color. This character depends on the color of the iris, the colored circle in the middle of the eye. The iris comprises two tissue layers, an inner one called the iris pigment epithelium and an outer one called the anterior iridial stroma. It is the density and cellular composition of the latter that mostly affects the color of the iris. The melanocyte cells of the anterior iridial stroma store melanin in organelles called melanosomes. White light entering the iris can absorb or reflect a spectrum of wavelengths, giving rise to the three common iris colors (blue, green-hazel, and brown) and their variations. Blue eyes contain minimal pigment levels and melanosome numbers; green-hazel eyes have moderate pigment levels and melanosome numbers; and brown eyes are the result of high melanin levels and melanosome numbers (Sturm & Frudakis, 2004). Textbook accounts often explain that a dominant allele B is responsible for brown color, whereas a recessive allele b is responsible for blue color. According to such accounts, parents with brown eyes can have children with blue eyes, but it is not possible for parents with blue eyes to have children with brown eyes (Figure 6.1). This pattern of inheritance was

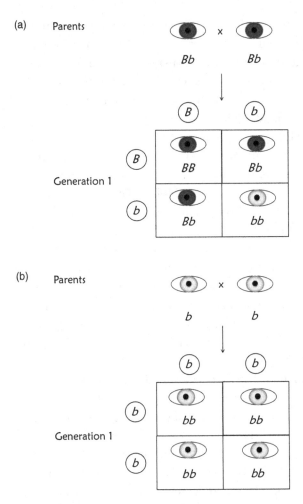

FIGURE 6.1 The typical account for the inheritance of eye color often found in textbooks. According to this account, a dominant allele determines brown color and a recessive allele determines blue color. Thus, parents with brown eyes can have children with brown or blue eyes (a), whereas parents with blue eyes can only have children with blue eyes (b) (brown eyes are shown in this figure as darker than blue eyes). However, this model of inheritance is insufficient to explain the observed variation.

first described in the beginning of the twentieth century (Davenport & Davenport, 1907; Hurst, 1908) and it is still taught at schools, although it became almost immediately evident that there were exceptions, such as that two parents with blue eyes could have offspring with brown or dark hazel eyes (e.g. Holmes & Loomis, 1909).

However, the inheritance of eye color, actually of iris pigmentation, is not as simple as presented in Figure 6.1. More than one gene has been found to be significantly associated with eye color. The strongest associations were initially found between eye color and the *OCA2* (OCA2 melanosomal transmembrane protein) gene, which is located on chromosome 15 and encodes a protein affecting melanosome maturation.[1] However, other strong associations between eye color and particular genes have also been found (Frudakis et al., 2003). Eye color is therefore best described as a polygenic character, i.e. one to which multiple genes contribute (Sturm & Frudakis, 2004). Nevertheless, it seems that particular genetic variants, closely located to one another, seem to account for blue eye color. It has been found that three SNPs within intron 1 of the *OCA2* gene have the highest statistical association with blue eye color (Duffy et al., 2007; see also Frudakis et al., 2007). Other studies have shown that SNPs in the introns of gene *HERC2*, also on chromosome 15, are strongly associated with blue color.[2] It has been assumed that the variants within the *HERC2* gene are related to the expression of *OCA2*, and that it is the decreased expression of the latter in iris melanocytes that is the cause of blue eye color (Eiberg et al., 2008; Kayser et al., 2008; Sturm et al., 2008; Sulem et al., 2007). Therefore, certain variants in *OCA2* and *HERC2* genes are significantly associated with blue eye color; however, it must be noted that these associations differ in different populations (Edwards et al., 2016). In this sense, one could say that there exist alleles "for" blue color, but these are not the alleles of a single gene. Furthermore, even if such a model worked in many cases, it would still be insufficient to explain the full range of eye color phenotypes that has been documented in research (Sturm & Larsson, 2009; see Figure 6.2). More generally, the association between a gene and a character is usually less than 100 percent. This means that the alleles of a certain gene can account for the phenotypic variation in a character, but cannot alone explain the whole variation observed.

A second important conclusion is drawn here: it is one thing to account for the variation of a character on the basis of particular genes,

---

[1] The gene is called *OCA2* because the respective protein is involved in the most common form of human occulo-cutaneous albinism, a disease characterized by fair pigmentation and susceptibility to skin cancer.

[2] The official name of *HERC2* gene is "HECT and RLD domain containing E3 ubiquitin protein ligase 2"(!).

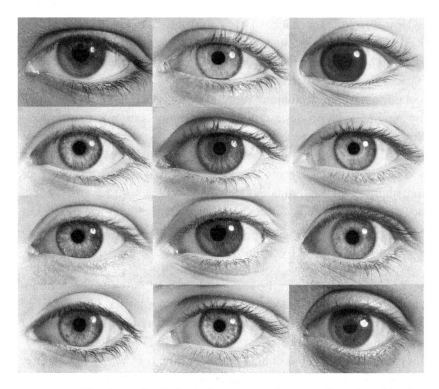

FIGURE 6.2 The observed variation in human eye color cannot be explained just by two alleles of a single gene (© Anthony Lee).

and another to causally explain how these genes produce the character. Whereas the former has been well understood in the case of eye color, there is still a lot to learn about the latter. This means that finding associations between genes and characters may be informative, but nevertheless does not reveal much about the underlying mechanism. We can find that a gene is related to a character, but this does not immediately show why and how the gene affects this character. Let me use a simple example to illustrate this distinction. Imagine that a math teacher in an international secondary school is coaching several student teams participating in math contests. After several years and several contests, the teacher decides to analyze the results and assess the performance of the various teams. The results indicate an interesting pattern: the higher the number of Asia-raised[3] students, the more

---

[3] Asia-raised refers here to people who were born, brought up and went to school in countries such as China, Japan, and Korea (not India!).

successful were the teams. This suggests that for some reason there is an association between being Asia-raised and success in math; thus, these students made a more significant contribution toward the success of their team compared to non-Asia-raised students. However, finding the association between having Asia-raised students and success in math is different from explaining why Asia-raised students achieve this. There could be several reasons for this, such as that Asia-raised students are smarter, or more mathematically inclined, or that they have developed a better attitude toward math compared to others because of language issues that make their counting system more logical. The latter actually seems a very plausible explanation as there is evidence that differences in mathematical competence appear around the age of four years old and reflect variations in number-naming systems that make accessing some mathematical relations more difficult in English than e.g. in Chinese (Miller et al., 1995; see also Gladwell, 2008, pp. 227–232). Therefore, the distinction between finding an association between two variables and explaining why this association exists is important. I return to this distinction in Chapter 8. For now, it is important to remember that the former does not entail the latter.

As already mentioned, researchers can study people having a disease and healthy people without that disease in order to see whether particular alleles are more common in one or the other group. More recently, this kind of study has been expanded beyond a few alleles and small, local populations to include thousands of participants and the search for numerous genetic variants (in the case of eye color, such a study was Kayser et al., 2008). These studies are called genome wide association studies (GWAS), and their focus is often on variations down to the level of a single nucleotide. GWAS use dense maps of SNPs that cover the human genome in order to look for differences in the frequencies of alleles between different groups of individuals. It is assumed that a significant difference in the frequency of a variant between two groups indicates that the corresponding region of the genome contains functional DNA sequences that somehow affect the respective character (or disease). It must be noted that in a GWA study, the ability to detect an association at a particular SNP decreases with the frequency of the less common variant at that SNP. In other words, the lower the frequency of the less common variant, the lower the ability to detect an association between that and the condition. This is why GWAS have detected associations with relatively more common variants (Donnelly, 2008). GWAS came to replace linkage studies, i.e. studies focusing on alleles shared by descent among relatives, as they have the advantage of having a higher sensitivity in detecting common variants with small effects, as well as that large samples of unrelated people can be collected much more easily than family-based samples (Kruglyak, 2008).

SNPs are not the only type of variants that can be found in DNA sequences. There also exist insertions-deletions, block substitutions, inversions of DNA sequences, and copy number differences (see Figure 6.3), which can be identified with various molecular methods (Feuk et al., 2006). However, these constitute only a small portion (between 0.5 percent and 1 percent) of any given genome, and therefore SNPs are the most prevalent type of genetic variation among individuals. Although rare and novel SNPs in individuals also exist, when the genomes of two individuals are compared, the majority of the nucleotides that differ are at positions of variants common in the whole population. These genetic variants contribute, along with environmental influences, to the development of characters. The full sequencing of human genomes has shown that in any given individual there are, on average, four million genetic variants.[4] The important task, of course, is to find out which of these variants actually affect phenotypes, and not just which ones are associated with them (Frazer et al., 2009).

The main principle behind GWAS is linkage-disequilibrium, or more simply the nonrandom association of alleles at two or more loci (for a detailed discussion see Slatkin, 2008). The idea is that a genetic variant, which somehow affects a character (or disease) and that is shared by some individuals because of common descent, will be surrounded by other shared genetic variants in a sequence that should correspond to that in the ancestral chromosome on which they first occurred. In other words, certain alleles are found close to one another not by chance but because they once were, and still are, linked and thus tend to be inherited together (these constitute a haplotype). The reason for this is that recombination through crossing over can rearrange alleles up to a certain point; some genetic variants are so close that they cannot be separated by recombination events. As a result, when researchers are

---

[4] Interestingly, but perhaps not surprisingly, the first two human genomes to be fully sequenced were those of Craig Venter (Levy et al., 2007) and James Watson (Wheeler et al., 2008). For an interesting account of these stories see chapter 1 of Davies (2010).

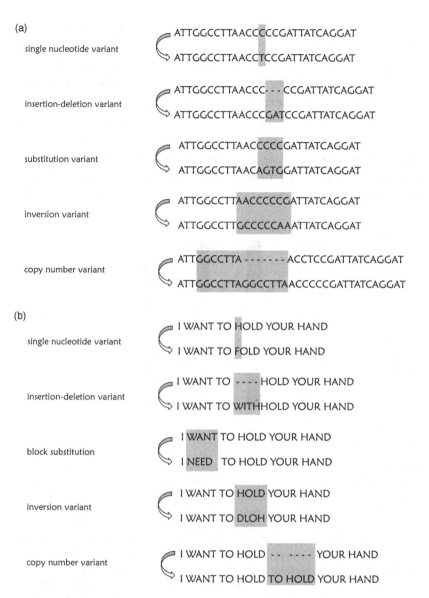

FIGURE 6.3 DNA variants in the genome. Single nucleotide variants are DNA sequence variations in which a single nucleotide (A, T, G, or C) is altered. Insertion-deletion variants occur when one or more bases are present in some genomes but absent in others. Block substitutions are cases in which several consecutive nucleotides vary between two genomes. An inversion variant is one in which the order of the base pairs is reversed. Finally, copy number variants occur when identical sequences are repeated and the difference is thus in how many times the repeated sequence exists in the genome.

looking for genes that are potentially associated with a character, they can first look for SNPs associated with it. When such SNPs are found, it is possible that a genetic variant somehow involved in the production of the character is located close to them. Then, a possible biological mechanism can be revealed (see Figure 6.4; see also Donnelly, 2008; Kruglyak, 2008). It must be noted that once the character of interest is associated with one or more SNPs, these are studied in independent samples for statistical validation. This is important because when thousands of SNPs are tested, many of them could be found to be statistically significant just by chance. For instance, if a study includes 500,000

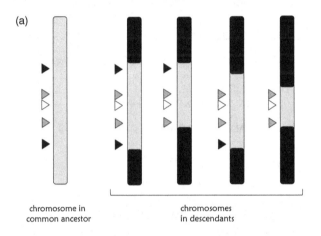

(a)

chromosome in
common ancestor

chromosomes
in descendants

FIGURE 6.4 (a) Linkage disequilibrium around a genetic variant, the position of which is indicated by the white triangle. This existed on the chromosome of a common ancestor, and was linked to some other genetic variants, the positions of which are indicated by the black and gray triangles. The ancestral chromosome in the common ancestor included all these variants, whereas the chromosomes of the descendants do not include all of them because some have been moved away due to consecutive recombination events, presented in (b) and (c) respectively (the part of each chromosome which is similar to the ancestral one is shown in gray, whereas the new parts introduced due to recombination events are shown in black). The genetic variants indicated by the black triangles may have been removed from some chromosomes because of recombination. In contrast, those variants indicated by the gray triangles are physically closer to the gene and so remain associated with it, despite the modifications brought about by recombination events. Therefore, these variants will be nonrandomly associated with the "white" genetic variant because they are linked. This property is used in GWAS as researchers can look for genetic variants and see if these are associated with a particular character. Once these are found, it is possible to also locate the gene associated with the character. Figure (a) is adapted by permission from Macmillan Publishers Ltd: *Nature Reviews Genetics* (Kruglyak, 2008), © 2008.

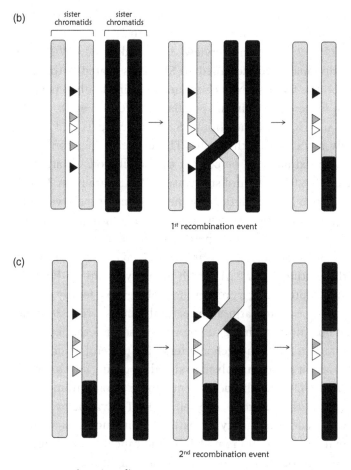

FIGURE 6.4 (continued)

SNPs, it is expected that 25,000 of them could be found to be statistically significant due to chance for an error of 0.05 (see Visscher, 2008; for a more detailed account of GWAS see McCarthy et al., 2008).

Let us consider a concrete example of the application of GWAS in the study of human height. The inheritance of this character was first discussed by Galton (1886), who argued for a clear pattern of resemblance among members of the same family. We now know from various studies that approximately 80 percent of the variation in height among individuals within a population is due to genetic factors (Visscher et al., 2006).[5] Several variants have been found to influence this character. In

---

[5] This is often related to the concept of heritability. This is a highly misunderstood concept that is discussed in detail in Chapter 10.

2008, 3 GWAS in a combined sample size of approximately 63,000 individuals concluded that 54 variants were found to be associated with height (see Visscher, 2008 for an overview). One study used GWAS data from 13,665 individuals and identified 20 SNPs associated with adult height. However, the twenty SNPs combined explained only 3 percent of the variation in height within the population (Weedon et al., 2008). Another study used GWAS data from 15,821 individuals and found 12 loci strongly associated with height. These twelve loci combined accounted for approximately 2 percent of the population variation in height (Lettre et al., 2008). A third study used GWAS data from 33,992 people, and identified 27 sites of the genome with one or more variants that showed significant association with height. Taken together, these variants explained approximately 3.7 percent of the population variation in height (Gudbjartsson et al., 2008). Variants related to four genes were found in all three studies: HMGA2 (high-mobility group AT-hook 2) on chromosome 12, producing a protein that belongs to a family of proteins that function as chromatin architectural factors; ZBTB38 (zinc finger and BTB domain containing 38) on chromosome 3 that encodes a transcription factor; HHIP (hedgehog interacting protein) on chromosome 4 that encodes a protein involved in signaling during developmental processes; and CDK6 (cyclin-dependent kinase 6) on chromosome 7 that is involved in the control of the cell cycle. The various variants identified are related to a variety of pathways, including signaling, extracellular matrix, chromatin remodeling, and cancer, indicating the complexity of the biological regulation of human height.

That so many variants explained so little variation in height (between 2 percent and 3.7 percent in the studies cited) might indicate that other, not-yet-identified SNPs with small effects might affect this character. It would also be possible that there exist variants that the GWAS approach does not capture. More recent studies have provided a more detailed picture. Whereas it is possible that unidentified genetic variants could account for the variation in height, it could also be the case that the actual effect of the identified ones was missed. Studies such as those described analyze GWAS data by testing each SNP individually for an association with a character. The statistical tests used reduce the occurrence of false associations, but may also miss real associations if the respective SNPs have a small effect. A different approach is to estimate the variance explained by all SNPs together. This was attempted in a study with GWAS data on 294,831 SNPs from 3,925 unrelated individuals. It was found that 45 percent of variance for height could be explained by considering all SNPs simultaneously (Yang et al., 2010). This was further confirmed by another study using GWAS data on 586,898 SNPs from 11,586 unrelated individuals. This study also showed that the variance explained by each chromosome is proportional to its length, and that SNPs in or near genes explain more variation than those between genes (Yang et al., 2011).

Apparently, numerous genetic variants contribute to human height. In a study that used data from 183,727 individuals, it was shown that hundreds of genetic variants, in at least 180 loci, were associated with variation in human height. These 180 loci were found to be related to 21 genes mutated in human syndromes characterized by abnormal skeletal growth. In addition, thirteen of these genes were closest to the most strongly associated variant. This data could explain approximately 10 percent of the phenotypic variation in height (Allen et al., 2010). A more recent study used GWAS data from 253,288 individuals and identified 697 SNPs in 423 loci that explained 16 percent of the observed variation. The main conclusion of this study was that human height is influenced by a very large number of variants. These are found throughout the genome, but are related to pathways clearly relevant to skeletal growth. This study also concluded that by conducting larger GWAS, it might be possible to identify SNPs that explain more of the observed variation in height (Wood et al., 2014).

Human eye color and human height are two examples that show that more than one gene is implicated in the development of characters. Whether these are just a few or hundreds is not important for our discussion here. What is important is that there is no single "gene for" eye color or height. How about behavioral characters? Are genes associated with certain behaviors? If yes, to what extent? It should be obvious to everyone that we do not inherit our DNA only from our parents. As everyone knows from experience, parents (biological or adoptive) usually prescribe the immediate social environment in which a person is brought up. Thus, children actually inherit from their parents this social environment, as well as several other relevant features including values, principles, and worldviews. Eventually, the behavior of children is shaped in this context, influenced by the behavior of their parents, as well as by several numerous factors that are directly related to this environment, such as schooling, friends, the wider family, and more.

Yet, it is the reference to genes that really seems to excite people when it comes to human behavior.

In 1965, a possible genetic link to aggressive behavior was reported in *Nature*. On the basis of previous research, the researchers assumed that if the presence of an extra Y chromosome predisposed its carriers to unusually aggressive behavior, then an increased frequency of XYY males among violent people should be expected. The researchers observed the chromosomes of "mentally sub-normal male patients with dangerous, violent or criminal propensities" in an institution. They studied the chromosomes of 197 people and found that 7 of them were XYY, a frequency that they considered much higher than what one should expect in a "normal" population. The researchers nevertheless concluded: "At present it is not clear whether the increased frequency of XYY males found in this institution is related to their aggressive behaviour or to their mental deficiency or to a combination of these factors" (Jacobs et al., 1965, p. 1352). In a study of a similar sample of 942 male patients, 7 were found to be XXYY. When the authors compared their findings with other studies, they found no significant differences in the frequencies of other abnormalities (e.g. XXY), and they concluded that what was different was the "extraordinary excess of individuals" with XXYY. They also reported that these people were taller than the others, making the inference that the additional Y chromosome had an effect on growth (Casey et al., 1966).

Let us first see how XYY males come to be. In Chapter 3, I presented the process of meiosis through which the gametes are produced. The gametes of a male should normally include either an X or a Y chromosome (Figure 3.5). However, mistakes are possible during this process. One of these mistakes is the nondisjunction of the two sister chromatids of the same chromosome during the second division. This results in the production of two gametes that have either one more or one less chromosome than the regular number (this condition is called aneuploidy). Thus, there can be a spermatozoon with twenty-two chromosomes, lacking a Y chromosome. There can also be a spermatozoon with twenty-four chromosomes and two Y chromosomes (Figure 6.5). If a gamete like the latter is combined during fertilization with an ovum with twenty-three chromosomes, the result will be an embryo with forty-seven, instead of forty-six chromosomes. As this embryo will have two Y chromosomes, it will be XYY (Griffiths et al., 2005, p. 493).

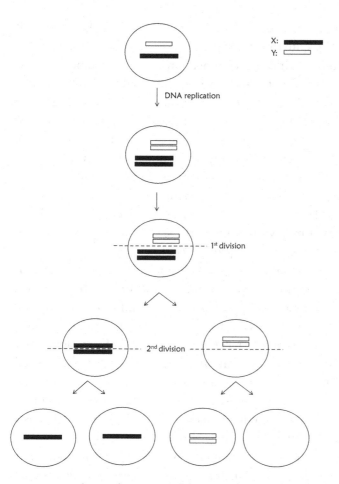

FIGURE 6.5 The nondisjunction of the sister chromatids of the Y chromosome may produce a gamete with two Y chromosomes, which in turn can produce an offspring with XYY after its fusion with an ovum with twenty-three chromosomes.

In the 1960s, several cases linked the XYY condition to criminality because of extensive publicity about genetic anomalies found in the body cells of certain criminals. This generated a lot of premature speculation and confusion (Fox, 1971). In one of these cases, Richard Speck (1941–1991) murdered eight student nurses in Chicago during one night in July 1966. Speck was tall, semiliterate, and had acne. This prompted some geneticists in 1967, after his trial and conviction, to conduct a chromosome analysis based on a sample from Speck's blood, and it was eventually reported that he was XYY. Despite a later study by cytologist Eric Engel (1925–2011) that showed Speck to be XY and not XYY, the press consistently described him as an XYY. Engel himself noted later on: "Despite the public release of repeated chromosome test results, magazines continued to comment on his chromosome anomaly, scientists still referred to it at meetings, and medical textbooks adopted it as a notable case" (Engel, 1972). Later studies explicitly discussed the limitations of the older ones, such as that reports were based on one or few cases, or that the search for XYY was done in selected groups, such as institutionalized men and tall men, and not in the general population. A major study that was conducted in Denmark, a country selected because of its excellent social records, involved 4,139 men and identified 12 XYY and 16 XXY men. The authors concluded: "No evidence has been found that men with either of these sex chromosome complements are especially aggressive" (Witkin et al., 1976).

In 1982, Patricia Jacobs (1934–), first author in the 1965 *Nature* article cited above, recounted what happened in the 1960s. She identified two problems. The first was "the enormous amount of publicity given to our findings in the lay press ... this resulted in a great many people, both scientific and lay alike, receiving their first information about the possible association of an additional Y chromosome and antisocial behavior from these sensationalist and untrue accounts in the media. In retrospect, I should not have used the words 'aggressive behavior' in the title of my paper and should not have described the institution as a place for 'the treatment of individuals with dangerous, violent or criminal propensities.'" The second problem "was the reaction, often made very publicly, of the large number of dedicated environmentalists who clearly felt threatened by the suggestion that there might be a genetic component to behavior. It was inconceivable to me and my colleagues that in 1965 there was a large body of professional people who did not and could not accept that both genes and environment played an integral role in every aspect of biology, including human behavior" (Jacobs, 1982, pp. 693–694).

Aggressive behavior attracted attention anew in the 1990s. In 1993, a study reported findings in a Dutch family where certain males exhibited mental retardation and aggressive behavior that was usually triggered by anger. This behavior could last for a few days, during which time these males slept very little and experienced frequent night terrors. This condition was found to be associated with a locus on the X chromosome, which seemed to be related to monoamine oxidase A (MAOA), an enzyme involved in the metabolism of neurotransmitters, such as dopamine, norepinephrine, and serotonin. Biochemical analysis indicated a disturbance of monoamine metabolism in the affected males who were tested. It was therefore concluded that a mutation affecting the *MAOA* gene might be responsible for the observed behavior (Brunner, Nelen, Van Zandvoort et al., 1993). In another study published in the same year, the researchers described this mutation in more detail. In each of the five affected males tested, a single nucleotide change was identified in the eighth exon of the *MAOA* gene, which resulted in the replacement of glutamine by a stop codon, and thus to an altered (shorter) MAOA enzyme. All affected males had a C to T substitution (i.e. C had been replaced by T) at a certain position, whereas twelve normal males of the same family had C at that position (Brunner, Nelen, Breakefield, et al., 1993). Interestingly, a report was published in *Science*, under the title "Evidence found for a possible aggression gene," although its author cautioned that it would be premature to generalize based on results from a single family, as well as that social, economic, and cultural factors might also be at work (Morell, 1993).

This finding attracted other researchers who attempted to study the possible connection between the *MAOA* gene and aggressive behavior. It was already known that a polymorphism in the promoter region of the *MAOA* gene affected its expression. The polymorphism consists of a repeat (repeated sequence) of 30 nucleotide pairs, present in 3, 3.5, 4, or 5 copies. The variants with 3.5 or 4 copies of the repeat were found to be transcribed up to 10 times more efficiently than those with 3 or 5 copies. The variants with the 3 and the 4 copies were also the most usually found ones, accounting for more than 95 percent of the observed variation (Sabol et al., 1998; Denney et al., 1999). Thus, researchers decided to look at these variants of the *MAOA* promoter, the corresponding MAOA expression and possible associations with different kinds of exposure to childhood maltreatment. In particular, researchers studied a cohort of approximately a thousand children, about half of which were male, and which were assessed at various ages until the age of twenty-six years old. For this sample it was known that between the ages of three and eleven years old, 8 percent of the children had experienced "severe" maltreatment, 28 percent had experienced "probable" maltreatment, and 64 percent had experienced no maltreatment. On the basis of the hypothesis that the *MAOA* genotype can moderate the influence of childhood maltreatment on neural systems implicated

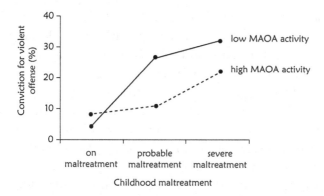

FIGURE 6.6 The association between childhood maltreatment and subsequent antisocial behavior as a function of MAOA activity, in the Caspi et al. (2002) study. The graph presents the percentage of males convicted of a violent crime by age twenty-six and childhood maltreatment in high and low MAOA activity individuals. The effect of maltreatment was found to be significant in the low MAOA activity group (163 individuals), and not significant in the high MAOA group (279 individuals). Therefore, the *MAOA* gene promoter genotype seemed to affect aggressive behavior (data from Caspi et al., 2002, p. 852, Figure 2b; the graph here is drawn with approximation).

in antisocial behavior, the researchers tested possible associations between variants in the promoter of the *MAOA* gene that conferred low or high MAOA expression, maltreatment, and antisocial behavior. It was concluded that severely maltreated children with the promoter variant conferring high levels of MAOA expression were less likely to develop antisocial behavior than severely maltreated children with the promoter variant conferring low levels of MAOA expression (Figure 6.6). The researchers noted that 85 percent of the latter had developed some form of antisocial behavior, implying that this was not a coincidence. Yet, they also noted that this study only provided preliminary findings and that further studies were required to see whether these results could be replicated and confirmed (Caspi et al., 2002).

Several studies attempted to replicate these results. A recent meta-analysis of twenty-seven published studies, from the original 2002 one until 2012, investigated the interaction of the *MAOA* genotype and childhood maltreatment in antisocial behavior. The meta-analysis concluded that across twenty male cohorts, early adversity was associated with antisocial outcomes more strongly for low MAOA activity individuals, compared to high activity ones. Therefore, this meta-analysis confirmed the findings of the initial study by Caspi et al. (2002). The authors concluded that maltreatment presumably interacts with MAOA to augment aggressive or antisocial potential, previously admitting that the biological mechanisms underlying the correlations under discussion remained unknown. Most importantly, they noted: "Of course, even a positive meta-analysis does not exhaust validity challenges or vouchsafe a true association" (Byrd & Manuck, 2014). Another meta-analysis of data for MAOA from thirty-one studies found a modest, positive association between the low activity MAOA variant and antisocial behavior (Ficks & Waldman, 2014).

In contrast, another meta-analysis of data from seventeen studies found no significant association between the *MAOA* gene promoter variants and aggression. The authors noted, interestingly, that this lack of association could be explained by the following considerations: (1) that a complex behavior like aggression is explained by hundreds or thousands of genes with complex interactions, rather than a few candidate genes; (2) that aggression and violence are complex behaviors that exist on a continuum, and so it is more likely that they are determined by many genes of moderate or small effect; and (3) that the sample sizes used in the reviewed studies were small, whereas GWAS studies show that very large samples are needed to reveal interesting findings in most human traits and diseases (Vassos et al., 2014). It must be noted that not only the *MAOA* variants but also others such as 5HTTLPR, the serotonin-transporter-linked polymorphic region of the serotonin transporter gene, have been considered as related to aggression. More broadly, most candidate gene association studies have identified associations between aggressive behavior and genes involved in neurotransmission and in hormone regulation, whereas GWAS have not yet identified genome-wide significant associations. Apparently, further studies are required, which would use larger samples and homogeneous phenotypes, in order to identify the genes underlying aggressive behavior. But aggressive behavior can be of different forms, such as increased anger, decreased fear, high emotional reactivity, or decreased control. Therefore, aggression in general is perhaps a vague description of a phenotype, and neuroimaging data could be used for quantifying the activity of certain brain regions associated with these different forms of aggression (Fernàndez-Castillo & Cormand, 2016).

So what does all this mean? Is the *MAOA* gene the "gene for" aggression or not? Apparently, not. It is already clear that several genes are implicated in the development of this complex character, and even clearer that further studies are required to clarify the underlying biological processes. But what is important to note for our purposes here is that even if it were the *MAOA* gene promoter variants alone that affected aggressive behavior, there would still be problems in deciding how to make good use of this information. Already in 2007, it was argued that even if the Caspi et al. (2002) results were replicated and confirmed, the conclusions would not be definitive. Most importantly, two important findings were noted: "childhood maltreatment is the strongest predictor of violent or antisocial behavior" and that "variation in the MAOA gene is not predictive for antisocial behaviors later in life" (Morris et al., 2007, p. 6). The first finding is evident in the graph in Figure 6.6. The amount of convictions for violent offense is less than 10 percent, both for the low MAOA activity and for the high MAOA activity individuals when they have not experienced any childhood maltreatment. In contrast, the amount of convictions approximately triples for either of these groups when individuals have experienced severe childhood treatment. Therefore, doesn't childhood maltreatment look like a more important factor for aggressive behavior than the *MAOA* alleles? The second finding is that it is not possible to predict which children with low MAOA expression will exhibit antisocial behavior in the future, as the association is probabilistic. Therefore, not only there is not a "gene for" aggression, but also even if it existed things would not be simple and straightforward.

In this chapter, starting from the characters of Mendel's peas, I focused on three human characters (eye color, height, and aggressive behavior) in order to show that several genes – in some cases a few, in other cases many – are implicated in their development. Therefore, the simple Mendelian model of inheritance that you learned at school and that was described in Chapter 1 does not really work for most characters. There are no "genes for" characters, but there generally exist several (a few or many) genes that somehow affect characters. The general conclusion from the particular characters discussed in this chapter is that differences in genes or other DNA sequences can be associated with differences in characters. Therefore, as Morgan and his colleagues explained more than 100 years ago (see Chapter 2), many genes should be considered for explaining the existence of a character; however, changes in a single gene might be sufficient to explain the difference between two versions of the same character (this is discussed in more detail in Chapter 10). Let us now see what happens in the case of disease. Perhaps "genes for" disease exist, even if there are no "genes for" characters?

7 Are There "Genes for" Diseases?

What is a genetic disease? Definitely something bad that we all wish to avoid. But does it differ from the characters discussed in the previous chapter? Disease cannot be defined in absolute terms, but only in relation to some state that we have agreed to consider as "normal." If we have agreed on some specific "normal" levels of hemoglobin, the molecule that transfers oxygen and carbon dioxide in blood, then lower levels resulting in insufficient transfer of these gases inside the body can be considered as a disease (one such case is β-thalassemia). Similarly, if we have agreed on some specific "normal" levels of cholesterol in blood, then higher levels that might in the long term be related to heart problems can be considered as a disease (one such case is familial hypercholesterolemia). But in both cases, these diseases emerge as deviations from what we consider as the "normal" state. Therefore, disease is a character difference. The conclusion of the previous chapter, that differences in genes cannot explain the existence of characters but can explain differences between different versions of a character, thus forms the basis for defining genetic disease. A factor considered to have caused the disease is one responsible for the transition from the "normal" state to the "disease" state. In both β-thalassemia and familial hypercholesterolemia the factor responsible is assumed to be a change in a particular gene. Beyond that, the approach to understanding the development of a disease is the same with the approach to understanding the development of a character: we usually first find correlations between particular DNA sequences and the occurrence of particular diseases, and then we try to figure out the pathway through which these DNA sequences may bring about the respective diseases (see Keller, 2010, pp. 45, 47).[1]

---

[1] It must be noted at this point that health and disease do not depend only on biological factors, but also on social and economic factors (Hubbard & Wald, 1997, p. 58). Whether one suffers from a genetic disease depends not only on one's genetic makeup or one's development in a particular environment, but also on one's lifestyle, nutrition, and access to healthcare. Nevertheless, the discussion of these important factors falls outside the scope of this book.

As explained in the previous chapter, variants that are somehow associated with characters can be identified through the GWAS. This is also the case for disease. However, especially in the case of disease, simply finding that some variants are associated with the disease is not enough to know; it is also important to know where these variants are located. Let us imagine that two variants of the same gene, V1 and V2, are found to be strongly associated with disease D. If these variants are on the same chromosome (they are "in phase," and this is why the process to map them is called "phasing"), the individuals who carry them will still have the "normal" allele on the other, homologous, chromosome. In the majority of diseases, this person will most likely not exhibit symptoms of the disease. However, if V1 and V2 are located each on one of the two chromosomes, then this person will be a compound heterozygote and will have two alleles that are malfunctioning or nonfunctional. Therefore, it is very likely that this person will exhibit the disease (Snyder, 2016, pp. 15–16). This simply entails that any attempt to sequence genomes and identify potential disease variants should go at a fine level of detail in order to figure out not only which variants exist but also exactly where they are located.

Let us examine in some detail β-thalassemia and familial hypercholesterolemia, which are usually cited as examples of a "recessive" and a "dominant" Mendelian disease, respectively. Hemoglobins are complex proteins contained in the human red blood cells (erythrocytes). All hemoglobins consist of a molecule (heme) that carries oxygen and four protein chains (globins) – hence the name hemoglobin. In all cases, α-globin chains combine with other globin chains to give hemoglobins HbA (α2β2) and HbA2 (α2δ2), which are the two main types in adults, as well as fetal hemoglobin HbF (α2γ2) which also exists in adults in very small amounts. The β-thalassemias occur when there is decreased or no production of β-globin chains and thus of HbA. This is due to mutations within, or related to the expression of, the *HBB* (hemoglobin subunit beta) gene on chromosome 11 that encodes the β-globin chains. Some mutations completely inactivate the *HBB* gene, resulting in no β-globin production and causing β0-thalassemia. Other mutations simply cause a reduction in the amount of β-globin produced, thus causing β+ thalassemia. This leads to an excess of α-globin chains, which aggregate in the precursor red blood cells causing their abnormal development and in mature red blood cells causing membrane damage and cell destruction, which in turn results in anemia that retards growth and development. Generally speaking, the disease has been considered to be a monogenic one, exhibiting a pattern of typical Mendelian inheritance as shown in Figure 7.1a (Weatherall, 2001).

Thalassemias are some of the most common genetic diseases worldwide. The first formal description of thalassemia was published in 1925, but the term was coined later (Cooley & Lee, 1925; Whipple & Bradford, 1936; for the relevant history see Loukopoulos, 2014; Weatherall, 2004). However, it was eventually realized that the phenotypes of β-thalassemias have an extremely high clinical heterogeneity. Generally speaking, there are three main clinical phenotypes of β-thalassemia: major, intermedia, and minor. On the one hand, many homozygotes and certain compound heterozygotes (i.e. people carrying two different alleles that are both associated with the disease) have β-thalassemia major and severe anemia from early in life, which may lead to death during the first year. Only if the symptoms are controlled by blood transfusion that provides the missing hemoglobin, and only if the excessive amounts of iron thus administered are removed, can children survive to adulthood. On the other hand, heterozygotes, i.e. those carrying one β-thalassemia allele, and other compound heterozygotes can have a variety of conditions ranging from β-thalassemia major and severe anemia to mild β-thalassemia intermedia without serious clinical symptoms. The main problem in β-thalassemia is not the underproduction of hemoglobin, but rather the excess amounts of α-globin chains. Therefore, people who have simultaneously developed some form of α-thalassemia along with β-thalassemia end up having less α-globin chains in excess. This is also the case in people with increased HbF production in whom γ-globin chains bind some of the excessive α ones. In cases like these, people have less severe anemia, although they also have less hemoglobin than "normal" (Weatherall, 2001; Higgs et al., 2012). This suggests that the model in Figure 7.1a does not work for all cases, as heterozygotes could have almost "normal" or totally "abnormal" levels of hemoglobin as shown in Figure 7.1b.

A GWAS, which involved 235 mildly and 383 severely affected patients, identified 23 SNPs in 3 independent genomic regions as being significantly associated with the severity of the disease. Not surprisingly, the highest association was found with SNPs within the *HBB* gene. The second most significant association was identified with two

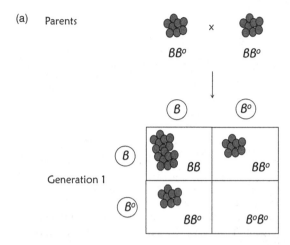

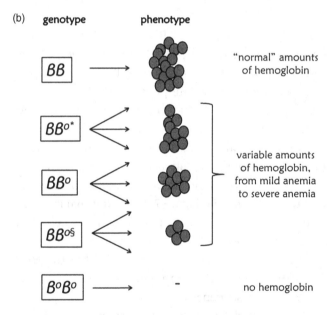

FIGURE 7.1 (a) The inheritance of β-thalassemia is usually described as that of a typical Mendelian character. According to this, two parents who are carriers of the defective allele have 25 percent probabilities of having an offspring who is homozygous and has the disease (circles represent the proportion of hemoglobin molecules; β^o is a defective allele and β is a "normal" one). (b) However, this model is insufficient to describe the phenotypic heterogeneity of the disease, as different heterozygotes can have a variety of phenotypes, from almost "normal" to almost defective ($\beta\beta^o$, $\beta\beta^{o'}$, $\beta\beta^{o\S}$ denote heterozygotes with different alleles).

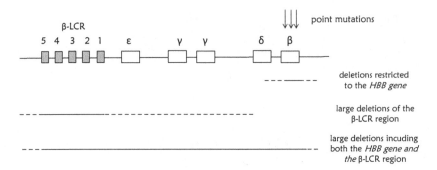

FIGURE 7.2 The mutations in the *HBB* gene and the respective locus control region (β-LCR) related to β-thalassemia. Mutations include point mutations that affect the β gene, deletions that are restricted to the β-gene, and large deletions involving the β-LCR with or without the β gene. The dashed lines represent variations in the amount of DNA removed by the deletions.

genes on chromosome 6, the *HBS1L* (HBS1 like translational GTPase) gene that encodes an enzyme, and the *MYB* (MYB proto-oncogene)[2] gene that encodes a protein that regulates transcription. Both of these proteins are involved in the processes of the production of blood cells. A third region was within the *BCL11A* (B-cell CLL/lymphoma 11A) gene on chromosome 2 that encodes a protein that seems to be implicated in leukemia (Nuinoon et al., 2010). Various molecular changes, including single nucleotide changes as well as more extensive changes such as deletions, have resulted in more than 300 alleles of the *HBB* gene that are related to β-thalassemia (Figure 7.2). Most β-thalassemias are due to point mutations in the *HBB* gene (see Giardine et al., 2014; see also Giardine et al., 2011). Several kinds of deletion are also associated with β-thalassemia. Some of them are restricted within the *HBB* gene (e.g. Thein et al., 1984). Other deletions are most extensive, removing part of this gene and other sequences, thus resulting in no production of β-globin chains. Other deletions remove regulatory sequences of the *HBB* gene, leaving the gene intact but resulting in its decreased expression (e.g. Kioussis et al., 1983). Thein (2013) provides a detailed description of the various types of more than 200 point mutations related to β-thalassemia in regulatory sequences, in sequences related

---

[2] Oncogenes are genes producing proteins that are related to cell division and proliferation. Mutations in these genes cause excessive cell division and proliferation that may result in cancer, and this is why they are called oncogenes.

to splicing, within the *HBB* gene sequence, as well as various deletions (see Figure 6.3 for a representation of these phenomena).

All the above point to an important distinction: it is one thing to have a mutation related to a disease and another to have the disease itself. Whereas patients suffering from β-thalassemia have one or more mutations related to the disease, there are many people who do not have the disease although they have a relevant mutation. Therefore, even in this classic and exemplar case of a monogenic disease there is heterogeneity that makes the notion of a "gene for" β-thalassemia difficult to sustain. This does not question the connection between the gene and the disease, as patients will most likely have one of the documented mutations. But one can question the inference made from a genetic test (especially a preimplantation or prenatal one)[3] finding such a mutation that the child-to-be will definitely suffer from the disease. Genetic counseling can be complicated when the clinical phenotype cannot be accurately predicted from the genotype, because – as already explained – some heterozygotes might have β-thalassemia major, and others with exactly the same genotype might have a very mild form of β-thalassemia intermedia (Higgs et al., 2012). Therefore, thinking of *HBB* as the "gene for" thalassemia is insufficient to give the whole picture.

Let us consider another disease, in this case a "dominant" one. Familial hypercholesterolemia is a condition that is characterized by high amounts of cholesterol in the blood, which results in coronary heart disease and heart attacks. The condition was first described on a genetic basis in 1938, as elevated cholesterol levels were found in families (Müller, 1938). The disease was soon found to be associated with a significant increase in the incidence of heart attacks. By the 1950s, it had been shown that fatal events increased in proportion to increased blood cholesterol levels. For example, comparative studies of Japanese men living in Japan, Hawaii, and California showed that fat intake,

---

[3] Preimplantation genetic tests can be performed in embryos formed during in vitro fertilization. A single cell of the developing embryo can be removed and be tested for the presence of alleles related to a disease. Then it can be decided whether the embryo should be transferred to the mother or not. Prenatal testing takes place during pregnancy by removing cells not from the embryo but by isolating cells of the embryo found in the fluid of the embryonic sac. These cells can then be tested with various methods, and decisions about whether or not to terminate the pregnancy can be made (see Braude et al., 2002; Sermon et al., 2004; Alfirevic et al., 2009; Harper & SenGupta, 2012; Bianchi, 2012).

plasma cholesterol, and the number of heart attacks all increased along the way from Japan toward the United States (Keys et al., 1958). Around that time, it also became clear that there is an association between blood lipoproteins and the risk for heart attacks (Gofman et al., 1950). It was soon shown that the plasma of people who had undergone a heart attack was characterized from increased levels of cholesterol-carrying lipoproteins – the low-density lipoproteins (LDL) – compared to individuals matched by age and sex without coronary heart disease (Gofman et al., 1955). Low-density lipoproteins consist of several hundreds of hydrophobic (nondissolving in water) molecules (cholesteryl ester) surrounded by a hydrophilic (dissolving in water) coat and a large protein called apolipoprotein B (apoB). LDL particles are considered to penetrate the walls of coronary arteries where they become oxidized, get inside white blood cells and convert them to cholesterol-laden foam cells. These initiate an inflammatory reaction, which makes the muscle cells of the arteries proliferate. In this way what is described as the atheromatic plaque grows (the phenomenon is described as atherosclerosis). At the same time, lipids derived from dead and dying cells accumulate in the central region of a plaque. When the plaque ruptures, a blood clot is formed that blocks the artery, thereby causing a heart attack (Libby et al., 2011).

During the 1970s, Joseph Goldstein (1940–) and Michael Brown (1941–) elucidated the molecular mechanism behind the increased LDL levels and familial hypercholesterolemia. They concluded that this disease results from mutations in the gene of the low-density lipoprotein receptor (LDLR) on chromosome 19. This is a trans-membrane protein that removes LDLs from the blood by binding to their apoB protein. Mutations in the *LDLR* gene result in the reduced uptake of LDLs, which in turn leads to the accumulation of LDLs in plasma, tendons, and skin, and thus contributes to atherosclerosis. At the same time, because of the limited or no LDL uptake, cells increase their own production of cholesterol by several times. Patients can be homozygous or heterozygous for this condition, and coronary heart disease occurs after a threshold of LDL exposure is reached, in early childhood in homozygotes and in early middle age in heterozygotes. The severity of the condition depends on whether the function of the LDL receptor is entirely disrupted or just reduced. The standard therapy for this condition is the administration of statins, which block cholesterol synthesis and thus lower cholesterol levels inside the cell, activate the synthesis of LDL receptors, and eventually lower the levels of LDL in plasma. Heterozygotes generally respond better to this kind of treatment than homozygotes, but statin treatment can be quite effective even in homozygotes if it starts before atherosclerosis develops. However, homozygotes are rather rare, with a frequency of less than one in a million people (Goldstein & Brown, 2015).

The *LDLR* gene encodes a protein that binds to the LDLs. More than 1,300 independent *LDLR* variants have been reported, more than three-fourths of which are due to base substitutions or small rearrangements in the exons. The rest of the changes include larger rearrangements, variants in the introns, and some promoter variants (Usifo et al., 2012). However, the *LDLR* gene is not the only one related to familial hypercholesterolemia. There is another gene, the *PCSK9* (proprotein convertase subtilisin/kexin type 9) gene on chromosome 1, which encodes a protein that is highly expressed in the liver and mediates the degradation of the LDL receptor molecules (Abifadel et al., 2003; Horton et al., 2009). The PCSK9 proteins bind to LDLRs in liver, promote their degradation and thus disrupt the process of recycling that returns them to the surface of the cell. Therefore, the number of LDLRs in the liver decreases and thus the amount of LDL in the blood rises (Zhang et al., 2007). However, it has been found that some people with certain variants of the *PCSK9* gene seem to be protected against heart attacks. In a study, from the 3,363 African Americans examined, 85 had nonsense mutations (ones that result in a stop codon and thus in a protein shorter than anticipated) in the *PCSK9* gene, which were associated with a 28 percent reduction in mean LDL cholesterol and an 88 percent reduction in the risk of coronary heart disease. Of the 9,524 European-Americans examined, 301 had a sequence variation in *PCSK9* that was associated with a 15 percent reduction in LDL cholesterol and a 47 percent reduction in the risk of coronary heart disease (Cohen et al., 2006). Therefore, particular mutations in *PCSK9* can be protective (Figure 7.3).

This is not the whole story, of course, as GWAS have identified associations between several other genes and coronary heart disease (Erbilgin et al., 2013). This suggests, again, that thinking in terms of a "gene for" hypercholesterolemia or coronary heart disease is insufficient to give the whole picture. As shown in Figure 7.4 two siblings with the same *LDL-R* alleles may have a very different phenotype because they carry

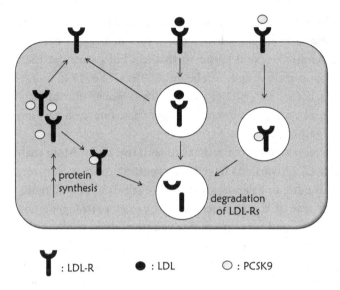

Y : LDL-R   ● : LDL   ○ : PCSK9

FIGURE 7.3 The role of LDL-R and PCSK9 proteins in LDL metabolism in a liver cell. The activity of LDLRs is the major determinant of plasma LDL concentration, as they bind and bring the LDLs within the cell. PCSK9 is a protein synthesized primarily in the liver that reduces the number of LDLRs in liver cells by binding to them and targeting them for degradation. Thus, as a result of PCSK9 activity, LDLR numbers are reduced and blood LDL levels increase. Gain-of-function mutations in the *PCSK9* gene produce a severe familial hypercholesterolemia phenotype, whereas loss-of-function mutations in this gene lower LDL-cholesterol levels by reducing LDLR degradation. Adapted by permission from Macmillan Publishers Ltd: *Nature Reviews Endocrinology* (Betteridge, 2013), © 2013.

different *PCSK9* alleles (Abifadel et al., 2009). It is thus clear that there is no single "gene for" familial hypercholesterolemia, and more than one gene should be taken into account for disease risk estimation.

Thalassemias are relatively common, and so is coronary heart disease, even if it is not always due to familial hypercholesterolemia. In these cases, interventions are possible in the form of blood transfusion in β-thalassemia or in the form of statin administration in familial hypercholesterolemia and coronary heart disease. However, there exist other types of genetic disease that are considered to be more influenced by the environment, and might not appear under certain conditions, although a person might carry the respective alleles. One such disease is phenylketonuria (PKU), a rare genetic disease that many physicians are taught about in medical school but very few actually encounter during their professional lives. It is a very interesting case because several decades ago it became the focus of newborn screening programs.

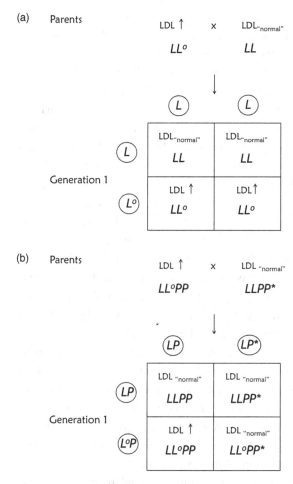

FIGURE 7.4 (a) When an individual with hyperlipidemia, who is heterozygous for a defective *LDLR* allele (*LL°*), and a "normal" individual have offspring, the probabilities for each of their children to have hyperlipidemia is 50 percent. All of their offspring with an *LL°* genotype would be expected to have high LDL levels in blood. (b) However, it is possible for an *LL°* individual not to have hyperlipidemia. This cannot be explained by the *LDLR* genotype, but only if one considers the effect of the *PCSK9* alleles. Individuals with *PP* genotypes can have problems. However, if a person carries an allele with a loss-of-function mutation in *PCSK9* gene (*P**), this allele can moderate the effect of the *LL°* genotype. Thus, this person does not have hyperlipidemia, as the LDL levels are within the normal range.

Screening refers to the search for conditions or specific alleles in the general, healthy, and asymptomatic population. The argument for the PKU newborn screening programs was that they would allow for early implementation of a special diet that would prevent the development of the disease. PKU thus became famous because its early identification was perceived as a victory of genetic medicine. The story is of course a bit more complicated (see Paul & Brosco, 2013).

PKU was first described in 1934 in two mentally retarded children. It was shown that their urine contained phenyl-pyruvic acid (Følling, 1934). On the basis of these and other observations, the disease – initially called oligophrenia phenylpyrouvica – was described as a new metabolic one, and was renamed phenylketonuria in order to link it to its characteristic metabolic by-product (Jervis, 1937; Penrose & Quastel, 1937). It was also shown that PKU exhibited the pattern of an autosomal recessive genetic disease, which was related to phenylalanine metabolism (Penrose, 1935; Følling, 1994; Centerwall & Centerwall, 2000). The disease was later attributed to a deficient activity of an enzyme (phenylalanine hydroxylase) that resulted in the inability to convert the amino acid phenylalanine to tyrosine, which in turn resulted in the accumulation of phenylalanine in the brain, causing severe neural impairment (Jervis, 1947, 1953). Reduced phenylalanine intake was also found to contribute to a less severe condition (Bickel et al., 1953).

Phenylalanine is an amino acid that humans need to acquire from food, as we cannot synthesize it (and this is why it is described as an essential amino acid). Once acquired, phenylalanine is either used in the construction of proteins or it is further converted to another amino acid, tyrosine. People with PKU cannot make this conversion, and so phenylalanine is accumulated in blood and tissues, and impairs the normal development of the brain (the exact process is not yet understood). PKU was considered, and is often described today, as a typical Mendelian recessive disease, which is caused by mutations in the *PAH* (phenylalanine hydroxylase) gene on chromosome 12, which in turn cause the impaired function of the respective enzyme, the excess amount of phenylalanine, (hyperphenylalaninemia) and mental retardation. A child who inherits two defective alleles, one from each parent, is expected to develop the disease because there is no active enzyme to metabolize phenylalanine (Figure 7.5a). PKU is also often described as an example of gene and environment interaction as people who follow a phenylalanine-restricted diet are supposed to be free from the disease.

However, it is not always possible to explain PKU as just the outcome of mutations in the *PAH* gene. Whereas such mutations are certainly sufficient to explain the phenotype at the enzyme level,

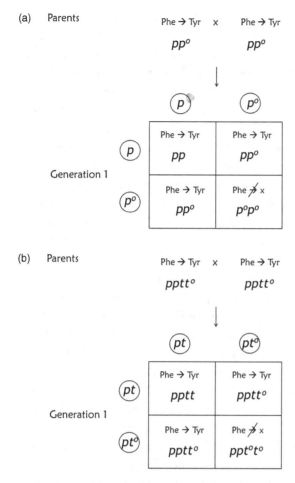

(a) Parents

Phe → Tyr × Phe → Tyr

pp^o pp^o

Generation 1

(b) Parents

Phe → Tyr × Phe → Tyr

$pptt^o$ $pptt^o$

Generation 1

FIGURE 7.5 (a) The inheritance of PKU can be described as a typical Mendelian character. According to this, two parents who are carriers of the defective allele have 25 percent probabilities of having an offspring who is homozygous and who has the disease (p^o is a defective allele and p is a "normal" one); (b) However, this model of monogenic disease is insufficient to explain hyperphenylalaninemia and the associated disease (PKU). There are several other factors affecting the phenotype, including other genes such as those that affect tetrahydrobiopterin recycling, which is an essential factor for the metabolism of phenylalanine. Thus, a person with "normal" genotype for the *PAH* gene may still have hyperphenylalaninemia because of mutations in a gene involved in tetrahydrobiopterin metabolism (indicated as *t* here).

they are not sufficient to explain the phenotypes at the metabolic (hyperphenylalaninemia) and cognitive (mental retardation) levels. Already in the 1970s, it was found that PKU could be not only due to mutations in *PAH*, but also due to problems in the metabolism of tetrahydrobiopterin, a protein essential for the catalytic function of phenylalanine hydroxylase (Smith & Loyd, 1974; Smith, Clayton & Wolf, 1975). Tetrahydrobiopterin deficiency can be caused by mutations in one of several genes,[4] which encode enzymes that are involved in the production and recycling of this protein in the body. Mutations that make one of these enzymes malfunction result in little or no tetrahydrobiopterin, which is necessary for the function of phenylalanine hydroxylase. Thus, hyperphenylalaninemia may occur without any mutations in the *PAH* gene and with "normal" levels of phenylalanine hydroxylase. However, mutations in the genes producing enzymes for the synthesis or recycling of tetrahydrobiopterin account for only 2 percent of patients with hyperphenylalaninemia (Thöny & Blau, 2006). But even in these rare cases, there is another gene besides the *PAH* that can account for PKU (Figure 7.5b).

Even more interestingly, and not surprisingly based on the other cases discussed in this chapter so far, it has been found that siblings with identical genotypes can have very different metabolic and cognitive phenotypes (e.g. DiSilvestre et al., 1991; Meli et al., 1998). It has also been found that phenotype-genotype correlations are not always clear. For example, in a meta-analysis of data from 365 patients carrying 73 different *PAH* mutations and exhibiting 161 different genotypes, it was found that 11 mutations were not consistent in their effects: 9 of them appeared in 2 different phenotype classes, and 2 appeared in 3 phenotype classes (Kayaalp et al., 1997). Overall, the PKU phenotypes exhibit complexities at all three levels: the cognitive, the metabolic, and the enzymatic. Several factors, besides mutations, seem to affect the disease phenotypes such as the transport of phenylalanine into the brain, the disposal of excess phenylalanine, or the degradation of mutant forms of the enzyme (Scriver & Waters, 1999).

All these findings suggest that PKU is not a simple monogenic disease, but a complex one. Therefore, what matters is the actual phenotype, and not the one predicted from the *PAH* genotype (Scriver, 2007). This is important, as there exist more than 850 *PAH* gene variants, which give rise to a continuum of conditions from very mild hyperphenylalaninemia, requiring no intervention, to severe PKU, requiring urgent

---

[4] Such genes are *GCH1* (GTP cyclohydrolase 1), *PCBD1* (pterin-4 alpha-carbinolamine dehydratase 1), *PTS* (6-pyruvoyltetrahydropterin synthase), and *QDPR* (quinoid dihydropteridine reductase).

intervention. As a result, most PKU patients are not homozygotes, but compound heterozygotes with two different alleles whose combination leads to clinically elevated phenylalanine levels. Approximately 50 percent of the mutations are missense ones, i.e. occurring due to nucleotide substitutions. There are other kinds of changes, such as mutations affecting RNA splicing, and small deletions. It should be noted that the position and the nature of the mutations seem to matter. Some mutations result in little or no enzyme activity and thus in the typical PKU phenotype; other mutations only partly inhibit enzyme activity, giving rise to mild phenylketonuria or mild hyperphenylalaninaemia; and about 5 percent of mutations do not affect the activity of the enzyme (Blau et al., 2010, 2014).

Therefore, it is no surprise that the implementation of phenylalanine-restricted diet does not always yield the anticipated results. As already mentioned, PKU became the first disease that could be identified through universal screening programs of newborns (Guthrie & Susi, 1963). This practice allows for the early diagnosis of the disease and the implementation of the phenylalanine-restricted diet, which is expected to help avoid the accumulation of excess phenylalanine in the body and thus the severe neural impairment. The implementation of the phenylalanine diet soon became a topic of controversy, as to whether it was always effective, whether it should be continued after the brain was supposedly developed, and above which phenylalanine levels the diet should be implemented. It seems that treatment for life that maintains phenylalanine within the recommended range is the most effective one for people with PKU, although many people seem not to have access to it (see Berry et al., 2013).

However, there are several cases in which diet-alone early treated PKU patients exhibit suboptimal outcomes. A systematic review of studies published since 2000 found 150 unique articles reporting PKU patient outcomes on diet-only therapy. In 140 of these articles there was at least one suboptimal outcome, including: neurocognitive/psychosocial outcomes such as intellectual, emotional, or behavioral functioning; quality of life outcomes such as difficulties in adhering to the strict diet; brain pathology outcomes such as white matter abnormalities, volume changes of grey at white matter, reduced cerebral protein synthesis, and altered cerebral metabolism; growth/nutrition outcomes such as deficiencies in several essential nutrients and micronutrients and increased body mass index; bone pathology such as imbalances in bone formation and decreases in bone mass density; and cases in which offspring could still be at risk for developing the symptoms because of untreated or late-treated maternal PKU (Enns et al., 2010).

Considering all these, PKU is more than a monogenic disease that can be treated with a special diet, as there are more than one gene involved. Even if most of the mutations are found in one of these genes, there is a range of phenotypes that depend on the kind of mutations that a person carries and thus phenylalanine-restricted diet does not always have the desired outcomes. Therefore, as in the case of thalassemia and familial hypercholesterolemia, a so-called monogenic disease is rather complex and cannot be explained on the basis of mutations on a single gene alone. I hope that this and the previous examples are adequate to illustrate the complexity of disease, and to make clear that there is really no single "gene for" any of these diseases, as well as that having the disease does not necessarily support the inference to the existence of a "gene for" it. The reason for this is not only that mutations in several other DNA sequences except for the main gene affect the disease, but also that a variety of disease phenotypes may exist even for people with the same genotype.

Things become even more complicated in the case of cancers. We usually talk about "cancer" in general, but this term implies that this is a single disease. However, this is very far from true as the term is used to describe a variety of diseases, which nevertheless have some common features. Cancers (also called neoplasms) result from the alteration of normal development and of tissue repairing processes during one's life. They can be solid tumors that develop in organs such as the breast, colon, and lung, or fluid tumors like those developing in blood during leukemia. A tumor is practically an overgrown mass of one's own cells. The problem they cause is that they consume so much of the resources of the body that the cells of the organ or tissue in which they develop die out, with bad consequences for the whole organism. When tumors are restricted to a tissue and grow slowly, they are not life-threatening, and they are called benign. In contrast, when they are invasive, moving from one tissue to another, they are life-threatening, and they are called malignant. In some cases, removal of the tissue is possible (e.g. of skin, colon, breasts) and if the whole tumor is removed then the individual can continue to live without problems. However, there are certain organs for such a removal is difficult or impossible (e.g. lungs, stomach, bone marrow), with the unfortunate outcome of death.

The vast majority of cancers are due to exposure to carcinogens either before or after birth, and they are described as sporadic. Carcinogens can be physical (e.g. α, γ, X-rays), chemical (e.g. tobacco, smoke, DDT, asbestos), or biological (e.g. viruses) factors. In principle, the development of cancers such as colorectal cancer, melanoma, and esophageal cancer are associated with exposure to environmental factors (see e.g. Wu et al., 2016). Inherited cancers are the minority of all clinical cancers (see e.g. Lichtenstein et al., 2000) and are due to mutations inherited from one's parents. Such cancers include retinoblastoma, breast cancer, ovarian cancer, and familial colon carcinomas. Therefore, it should be clear that there are no "genes for" cancer, although it is certainly the case that certain genes are associated with certain cancers (Sonnenschein & Soto, 2013). But, as I explain in Chapter 8, associations and causal connections are different.

Research overall suggests that tumor development in humans comprises multiple steps, each of which is characterized by alterations that progressively transform healthy human cells into malignant ones. Overall, it seems that most cancers exhibit six important physiological changes, which are acquired during the development of the tumor and which allow cells to overcome certain anticancer defense mechanisms that they naturally exhibit. These have been described as the hallmarks of cancer, and they are: (1) self-sufficiency in growth signals; (2) insensitivity to antigrowth signals; (3) evasion of programmed cell death (apoptosis); (4) limitless replicative potential; (5) sustained angiogenesis; and (6) tissue invasion and metastasis (Hanahan & Weinberg, 2000). Collectively, all these contribute to the uncontrolled proliferation of cancer cells and the development of the tumor. More recent research has revealed two more emerging hallmarks that have the potential to be as important as the other six: reprogramming of energy metabolism and evading immune destruction (Hanahan & Weinberg, 2011). Therefore, despite the differences in the more than 100 types of cancers that have been described, they are considered to share a common pathogenesis.

Normal human cells become cancerous ones after many of the aforementioned hallmarks occur successively, and only late-stage cancers exhibit all of them. This is the outcome of successive mutations that happen in the same or neighboring cells, which eventually become

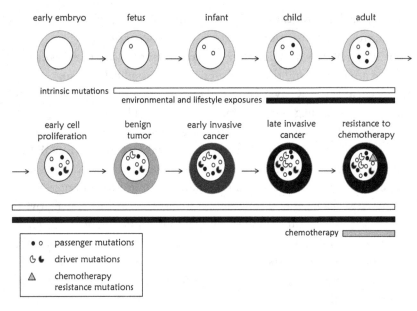

FIGURE 7.6 The timing of the somatic mutations acquired by a single cell during the development of cancer. Passenger mutations are those that do not affect the cancer cell, and driver mutations are those that cause proliferation. Furthermore, resistance mutations can occur that are associated with the reemergence of cancer after chemotherapy. Adapted by permission from Macmillan Publishers Ltd: *Nature Reviews Endocrinology* (Stratton et al., 2009), © 2009.

cancerous ones. Each time a certain hallmark appears, it becomes more likely that another one will occur next because cells become more disorganized. Thus, by continuously having more of such disorganized cells, it becomes more likely that further disorganization will take place. Worse than that, certain mutations already exist in some people – and this is why some types of cancer run in families. Figure 7.6 represents the lineage of cell divisions from the fertilized egg to a single cell within a cancer, showing the timing of the somatic mutations acquired by the cancer cell and the processes that contribute to them. Overall, cancers are the outcome of the accumulation of several mutations acquired by somatic cells.

Cancer development is, in fact, a process analogous to natural selection; it is based on the continuous acquisition of mutations in individual cells and the subsequent selection process that takes place among the phenotypically different cells. The first process results in what are described as somatic mutations, i.e. mutations occurring anew in the body. These mutations include all kinds of DNA sequence changes described in Figure 6.3. The rate at which such mutations occur increases in the presence of carcinogens. Cells may also acquire DNA sequences from exogenous sources such as viruses. Other processes, such as problems in the repair mechanisms of DNA, may further increase the number of newly occurring mutations. Among the genetically diverse cells thus produced, a selection process may take place. Cells with deleterious mutations may die out; but cells carrying mutations that confer the ability to proliferate and survive more effectively than their neighboring cells may also increase in numbers. Some of these cells may just remain invisible or develop into benign structures such as skin moles, but occasionally some may become cancer cells that are able to proliferate autonomously and invade other tissues (Stratton et al., 2009).

The way cancers develop point to a very important distinction, that between a genetic disease and an inherited disease. Whereas an inherited disease is quite often genetic as well, there exist numerous genetic diseases that are not inherited. Cancers are an exemplar case in which there may be a genetic change, i.e. a mutation in DNA, which was not inherited from one's parents (not a germline mutation), but which happened anew (de novo) in one's somatic cells (and this is why it is called a somatic mutation). Even in the cases of cancers that run in families, it is not always the case that they are due to genetic inheritance. In contrast, it is possible for family members to have mutations leading to cancer because they were exposed to the same carcinogen. We tend to think (and also to be taught at school occasionally) that we inherit our DNA from our parents in some fixed form. However, it has been shown that offspring often carry new mutations that do not exist in their parents and that are therefore novel. Using different methods, studies have estimated the mutation rate in humans to be 1.2×10^{-8} per nucleotide per generation (Kong et al., 2012), and 1.4–2.3×10^{-8} mutations per nucleotide per generation (Sun et al., 2012). Details notwithstanding, this data suggests that any individual is likely to carry at least seventy-two new mutations ($1.2 \times 10^{-8} \times 6 \times 10^9 = 72$).[5] This is why a genetic disease may not be due to some mutation inherited from one's parents but due to a new one.

Cancers are not the only disease of this kind. Another one is achondroplasia, a form of short-limb dwarfism in humans that affect

---

[5] The human genome is estimated to consist of 6×10^9 (six billion) nucleotides.

thousands of people all over the world. In the majority of cases (more than 95 percent), the disease is due to the same single nucleotide mutation in the *FGFR3* (fibroblast growth factor receptor 3) gene on chromosome 4. However, more than 80 percent of these cases are due to new mutations, most often in the father. This means that children with achondroplasia in four out of five cases do not have parents with the disease (Horton et al., 2007). Therefore, achondroplasia is not really an inherited disease, although it is certainly a genetic one, and one that makes the distinction between the two types clear.

Returning to cancers, it is very important to keep in mind that when mutations are found in tumors it is not clear which ones had a causal influence on the development of cancer (driver mutations) and which ones were irrelevant because the cancer had already developed (passenger mutations). It has been found that most somatic mutations in cancer cells are likely to be passenger mutations; however, about one in five genes (120 of 518) are estimated to carry a driver mutation and therefore may function as genes promoting cancer (Greenman et al., 2007). Driver mutations are causally implicated in the development of a tumor (oncogenesis) by conferring a growth advantage on the cancer cell that is then positively selected, i.e. proliferates faster than its neighboring cells. In contrast, passenger mutations do not confer any growth advantage, are not selected, and thus do not contribute to the development of cancer. They exist only because they just happened to occur. Apparently, distinguishing between driver and passenger mutations is a major challenge. As driver mutations are usually found in certain "cancer" genes,[6] whereas passenger mutations are more or less randomly distributed in the genome, it has been possible to identify many somatically mutated cancer genes in studies focusing on small regions of the genome. However, doing this for the whole genome is a lot more difficult (Stratton et al., 2009).

A very informative review (Vogelstein et al., 2013) of recent research in cancer genomics provides some important conclusions about the development of cancer. Overall, it seems that most human cancers are caused by two to eight sequential mutations that take place over the course of 20–30 years. Each of these mutations provides, directly or indirectly, a

---

[6] These are the proto-oncogenes that generally promote cell division and the tumor suppressor genes that generally inhibit cell divisions. Mutations in proto-oncogenes change them to oncogenes and result in a gain of function, thus enhancing cell division; mutations in tumor suppressor genes result in a loss of function, thus also enhancing cell division (see Varmus, 2006).

selective growth advantage to the cell in which it exists. Evidence suggests that there are approximately 140 genes in which mutations that contribute to cancer occur. These genes affect several signaling pathways that regulate three important cellular processes: cell fate determination, cell survival, and genome maintenance. The pathways affected in different tumors are similar, but the mutations in each individual tumor are different, even between tumors of the same type. As a result of the genetic heterogeneity of cells of the same tumor, there may be different responses to cancer therapy. Therefore, knowing the cancer genome of a given patient may be useful for the appropriate therapeutic plan, as well as for the early detection of cancer and the prevention of its development.

Given this, an important question is when and how often such mutations occur. Apparently, there are different processes generating mutations, but some of them seem to operate at a constant rate in all individuals resulting in a number of mutations in a cell proportional to the age of the person. A study of mutations from 10,250 cancer genomes across 36 cancer types found signs of clock-like mutational processes, i.e. processes producing mutations at a constant rate, in human cells. It seems that the rate of cell proliferation may be one important factor influencing the mutation rate; however, the underlying biological processes are not understood (Alexandrov et al., 2015). Another recent study supports this idea, suggesting that the lifetime risk of several types of cancer is strongly correlated with the total number of cell divisions that normally renew a given tissue. This means that many cancers develop in certain tissues because of random mutations occurring during DNA replication in normal cells and not because of external, carcinogenic factors.[7] This explains why cancers develop in certain human tissues a lot more often than in others: two-thirds of this variability is explained by newly occurring mutations as the more the cells in a tissue that divide, the more likely it is for mutations to occur and thus for cancer to develop (Tomasetti & Vogelstein, 2015).[8]

In this chapter, I have made a brief presentation of the complexities in the development of disease. Several genes may contribute to the

---

[7] As this article caused some debate, it must be noted that it does not claim that two-thirds of all cancers are due to random mutations and bad luck; it is the variability among tissues in developing cancer that is explained in this way (see Couzin-Frankel, 2015a, 2015b).

[8] Cancers can be due to not only genetic but also epigenetic changes (see Chapter 11).

development of diseases such as β-thalassemia, familial hypercholes-terolemia, and phenylketonuria. Environmental factors may also be important in these cases, and they are definitely important in coronary heart disease and cancers. The most important (and perhaps the most surprising) conclusion of GWAS has been that in most cases several genes were found to be associated with the risk of developing a disease. The relationship between genes and disease is usually a many-to-many one, as many genes may be implicated in the same disease, and the same gene may be implicated in several different diseases. At the same time, some alleles may be protective for one disease but increase the risk for another. For example, a variation in the *PTPN22* (protein tyrosine phos-phatase, nonreceptor type 22) gene on chromosome 1 seems to protect against Crohn's disease but to predispose to autoimmune diseases. In other cases, certain variants are associated with more than one disease, such as the *JAZF1* (JAZF1 zinc finger 1) gene on chromosome 7 that is implicated in prostate cancer and in type 2 diabetes (Frazer et al., 2009). Therefore, we should forget the simple scheme gene 1 → disease 1; gene 2 → disease 2; and adopt a richer – and certainly more complicated – representation of the relationship between genes and disease, such as the one presented in Figure 7.7. Additional GWAS on more variants in larger populations might provide a better picture in the future.[9]

The non-straightforward relation between genes and disease becomes more and clearer with more recent evidence, for instance in the case of prion diseases. These are rare neurodegenerative disorders caused by prion proteins that are misfolded. Approximately 63 mutations in the gene *PRNP* (prion protein) on chromosome 20 have been linked to them, and some are found relatively frequently. In sequence data from 16,025 invididuals with prion disease, the D178N mutation was found in 209 people, the V210I one was found in 247 people, and the M232R one was found in 63 people. The general assumption until recently had been that if one has one of these mutations, one will also get the dis-ease. In order to better understand this connection, however, research-ers looked into data compiled by the Exome Aggregation Consortium

---

[9] Some people have been skeptical about this prospect, questioning the value of studying more SNPs at larger populations with this joke: "At Harvard, you learn less and less about more and more until you know absolutely nothing about everything. At MIT you learn more and more about less and less until you know everything about nothing" (Davies, 2010, pp. 62–63).

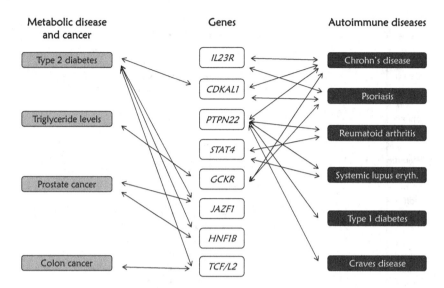

FIGURE 7.7 Co-contribution of several genes to human disease. Adapted by permission from Macmillan Publishers Ltd.: *Nature Reviews Genetics* (Frazer et al., 2009) © 2009.

(ExAC),[10] which contains the protein-coding sequences of 60,706 individuals representing various ethnic groups. As the total prion disease occurrence is 2 in every 1,000,000 people per year, one would expect to find less than two individuals – actually 1.7 – with *PRNP* mutations in that sample. However, 52 such individuals were found. Ten people had mutation M232R, suggesting that it probably doesn't cause disease or that it raises risk only marginally, 2 people had mutation V210I, suggesting that it may raise the risk only slightly, and no-one in the database had mutation D178N, suggesting that it is highly likely to have a strong causal, lethal effect (Hayden 2016; see also Lek et al., 2016; Minikel et al. 2016). These results show clearly that there is still a long way to go in order to understand the genetic basis of disease.

Beyond that, we should also keep in mind that in the case of complex diseases such as cancers, conclusions are difficult to make because finding associations is not necessarily informative. For instance, associations with cancer risk or benefits have been found for most food

[10] This is a coalition of researchers seeking to aggregate and harmonize exome sequencing data from various large-scale sequencing projects focusing on the exons of genes, and to make it widely available.

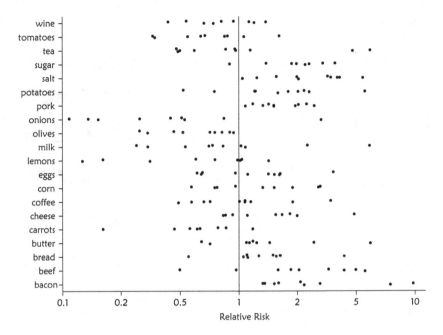

FIGURE 7.8 Effect estimates reported in the literature by food ingredient (only ingredients with ten or more studies are shown). Individual studies point to contradictory conclusions about the relation between food ingredients and cancer. This is why one should look at meta-analyses and many studies together, rather than individual ones. Reprinted by permission *Am J Clin Nutr* (2013; 97:127–134), American Society for Nutrition.

ingredients (Figure 7.8). A study aimed at examining the conclusions, statistical significance, and reproducibility of published studies on associations between specific foods and cancer risk. Fifty common food ingredients were selected, and articles about their cancer risk were found for forty of them. From 264 single studies, 103 studies concluded that the ingredient tested was associated with an increased risk and 88 studies concluded that the ingredient tested was associated with a decreased risk. However, thirty-six meta-analyses presented different results, with only four of them reporting an increased risk and nine of them reporting a decreased risk. It was thus concluded that multiple studies are necessary in order for scientists to draw valid conclusions, and caution is required when drawing conclusions until strong evidence becomes available (Schoenfeld & Ioannidis, 2013).

Technology in genomics may advance further, and analytical tools may be developed to such an extent that researchers might eventually identify every single variant and its contribution to one or more characters or diseases. Nonetheless, this might not be enough if the results of such studies are not appropriately understood and communicated. We have so far seen in some detail some characters and diseases about which it is clearly difficult to say that there exist particular "genes for" them. In all cases, there are some (a few or numerous) variants that are associated with particular versions of the character or the disease. However, and this is the most interesting part, the exact causal role of these variants is in many cases unclear. It is one thing to find that A is associated with B and another to explain why and how A is implicated in the production of B. I discuss this important distinction in the next chapter. The main problem for the public understanding of genetics and genomics is not only that this distinction is not always made, but also that the results of research are exaggerated in various ways. This may happen inadvertently, due to misunderstanding, because exaggerating may attract publicity, or because someone was looking for data to support an already reached conclusion. Let us now look into this important distinction.

8   So, What Do Genes Do?

The first conclusion from the last two chapters is that the common distinction between "simple" and "complex" or "monogenic" and "multifactorial" characters or disease is an oversimplification. Whereas it is certainly the case that some characters and some diseases are strongly associated with one gene, other DNA sequences may also be involved in their development. In addition, it is possible that within the main gene numerous variants exist that have a variety of phenotypic outcomes. Therefore, even in the case of "simple" or "monogenic" characters and disease, the path from the DNA sequence to the associated phenotype can be quite complex. As the examples in the previous chapters have shown, there are several reasons for this, such as that variation within a single gene can result in very different phenotypes, that phenotypes cannot be accurately predicted from genotypes even in cases of monogenic characters and disease, and that several genetic and nongenetic factors may also have a contribution to the respective phenotypes. Therefore, even monogenic characters and disease had better be considered as complex ones, too, even if they are less complex than others. This conclusion makes the idea of a single "gene for" a character or disease untenable.

But even at a theoretical level, the idea of "genes for" makes no sense conceptually. Moss (2003, p. 1) makes a very good point that should be obvious and intuitive, yet it is not often discussed. "Gene for" talk is about the attribution of characters and diseases to DNA, even though it is not DNA that is directly responsible for them. As shown clearly in the two previous chapters, it is proteins of different kinds that directly affect characters or diseases.[1] Therefore, it would make more sense to state that a person has a certain (defective) hemoglobin for β-thalassemia or low-density lipoprotein (LDL) receptor for familial

---

[1] Interestingly, the term "protein" was first used by Jöns Jacob Berzelius (1779–1848) and is derived from the Greek word "πρωτεῖος" ("πρῶτος" in modern Greek), which means first in rank or position (Vickery, 1950). It seems that we had to unravel the secrets of DNA in order to eventually realize, and perhaps reaffirm, the importance of proteins.

hypercholesterolemia or phenylalanine hydroxylase for hyperphenylal-aninemia. However, in all these and other cases we prefer to attribute a condition to a gene rather to a protein. Another important point is that several enzymes and other proteins are involved in the synthesis of any protein in our cells. As each of these enzymes and proteins is synthe-sized on the basis of some genes, all these are relevant to the synthesis of e.g. β-globin and not just the *HBB* gene that encodes it (Hubbard & Wald, 1997, p. 52). Therefore, if many genes produce or affect the pro-duction of a protein that in turn affects a character or a disease, it makes no sense to identify one gene as the gene responsible "for" this charac-ter or disease. Single genes do not produce characters or disease, and it is necessary to provide students and the public with a more accurate and more authentic view of what genes can and cannot do.

A first step in this direction would be to portray the actual poten-tial of genetics research in science education and communication. Until researchers are able to understand a biological process and how a change in one or more DNA sequences affects a character or the onset of a disease, all we can have are statistical associations. This became evident several years ago, as soon as the first results of DTCG tests became widely known. Several people tried, for different rea-sons, the results of the so-called $1000 genome analysis. Such exam-ples include journalists Nic Fleming and Boonsri Dickinson,[2] and genetics experts such as Francis Collins (Collins, 2010, pp. xvii–xxiv) and Kevin Davies who was the founding editor of the journal *Nature Genetics* (Davies, 2010, p. 149). They all had their DNA analyzed in the early days of DTCG testing, each one of them by three different companies (Fleming by GeneticHealth, deCODEme, and 23andMe, and all the others by Navigenics, deCODEme, and 23andMe). The results and the estimated risk that each of them received from the three companies were not always the same, and in some cases they even were contradictory. Around the same time, the results of 23andMe and Navigenics on thirteen diseases for five individuals were compared, and the same conclusion was reached: the different companies provided individuals with different risks for the same dis-ease (Ng et al., 2009; see Table 8.1).

---

[2] For these stories, see www.thesundaytimes.co.uk/sto/news/uknews/article234529.ece and http://discovermagazine.com/2008/sep/20-how-much-can-you-learn-from-a-home-dna-test, respectively.

TABLE 8.1 *Results from the Estimation of Relative Risk (RR) for Certain Diseases for Five Individuals*

Disease	Individual A (female)	Individual B (female)	Individual C (female)	Individual D (male)	Individual E (male)
Breast cancer	↑↑	↑↑	↓↓		
Celiac disease	↓↓	↓↓	↓↓	↓↓	↓↓
Colon cancer	==	==	=↓	↑↑	=↓
Crohn disease	↓↑	↓↑	↓↓	↓↓	↓=
Heart attack	↓↓	=↓	=↓	=↓	↑↑
Lupus	↑↓	↓↓	↓↓	↑=	↑=
Macular degeneration	↓↓	↓↓	↑=	↓↓	↓↓
Multiple sclerosis	↑↑		↓↓	↓↓	↓↓
Prostate cancer				↑↑	↓↑
Psoriasis	↓↑		↑↓	↑↑	↓↓
Restless legs syndrome	=↓	↑↑	↓=	↓↑	↑↑
Rheumatoid arthritis	↑↑	↑↑	↓↓	↓↓	↑↑
Type 2 diabetes	↓↓	=↓	↓↓	↑↓	=↓

Note: Symbols indicate the risk to develop a disease: ↑ increased risk (RR > 1.05); ↓ decreased risk (RR < 0.95); = equals average risk ($0.95 \leq RR \leq 1.05$). In all cases, the first estimation is from 23andMe and the second estimation is from Navigenics. Those cases in which the two companies gave different estimations for the same disease are highlighted in gray color.
Source: Data based on Ng et al. (2009).

There are two types of risk: absolute and relative. Absolute risk is the probability for a person to develop a disease over a period of time, usually a lifetime (in this case it is described as a lifetime risk). For example, an absolute risk of 0.5 indicates that the probability for a person to develop the disease over a period is 50 percent. In contrast, relative risk is defined with some point of reference in mind. For instance, if one suggests that the relative risk for a person carrying a certain genetic variant to develop a certain form of cancer is 1.5, the probability of this person to develop the particular disease is 50 percent higher than that of a person not having the same variant (Davies, 2010, p. 59). I discuss risk in more detail in Chapter 12; however, it is important to ensure for now that these numbers make sense. Psychologist Gerd Gigerenzer (1947–) has shown convincingly that numbers like this make little sense to people, and he has suggested to use frequencies rather than probabilities when talking about risks (Gigerenzer, 2002, chapter 4). An absolute risk of 0.5, that indicates that the probability for a person to develop the disease is 50 percent, practically means that 50 out of 100 people of a certain group will develop the disease over a certain period of time. Let us assume that this is a group of people who do not have a certain allele A. Imagine now that in another group of people carrying the allele A, the absolute risk of developing the disease is 0.75. This means that 75 percent, or 75 out of 100 people, of this group will develop the disease. In this case, the allele A is considered to increase the absolute risk for this second group. In other words, the increase of the absolute risk that the allele A brings about is 25 percent. Now the relative risk is the ratio of the probability of people carrying the allele A to develop the disease over the probability of people not carrying the allele A to develop the disease. In this case, this would be 0.75 / 0.50 = 1.5. This practically means that people carrying the allele have 50 percent higher probability to develop it than those who do not carry it.[3]

Now, how can the different estimates in Table 8.1 be explained? This largely depends on the set of variants that each company tests for in order to calculate the relative risk for each individual. As already explained, the findings of GWAS point to associations between certain variants and certain diseases. However, the various companies can be rather selective in which variants they analyze for a certain disease. This results in contradictory estimations of risk in developing a certain disease such as those presented in Table 8.1. Differences can also occur in the estimations of the same company across time, if along time new associated variants are found and are included in the tests. Thus, the estimated probabilities for each disease for each individual in Table 8.1

---

[3] I must note that the risks described in this example are very high, so they should not be taken at face value.

might be different today than in 2009, given the currently available data and that available in 2009. But this is not the most important point. What is important is for people to understand the probabilistic nature of these tests and of all medical testing. Relative risk for coronary heart disease may be estimated on the basis of genetic variants, but also on the basis of LDL levels as explained in Chapter 7. What is important to understand is that people with certain genetic variants and with higher LDL levels are more likely to develop coronary heart disease. The case is clearer for LDL levels because their impact on coronary heart disease has been studied for decades and, most importantly, because the underlying biological mechanism is now known. But even in this case, the connection is always probabilistic. Some of those with high LDL levels may develop coronary artery disease relatively early in life whereas others many not. Probabilities only indicate how likely an event is to happen, not whether or not it will happen.

A more recent study compared the methods of 23andMe, deCODEme, and Navigenics through a simulation. The study was conducted in a hypothetical population of 100,000 individuals on the basis of published genotype frequencies. The risks for developing particular diseases were calculated using the methods of the companies, as described on their websites. These diseases were age-related macular degeneration, atrial fibrillation, celiac disease, Crohn disease, prostate cancer, and type 2 diabetes, and they were chosen because they differ in the effect of SNPs and in the average population risks. Age-related macular degeneration and celiac disease are influenced by a few SNPs with strong effects on disease risk, whereas the other diseases are influenced by many SNPs with relatively weak effects. In addition, celiac disease and Crohn disease are rare disorders, whereas the others are more common. The results of this study showed a substantial difference in the predicted risks of the three companies because of differences in the sets of the SNPs used and in the average population risks they had chosen (Table 8.2), as well as in the formulas used for the calculation of risks. Given the choices made by the companies, as presented in Table 8.2, the differences in the estimated risks in Table 8.3 should not be surprising. These findings show that the results from DTCG tests should be considered with caution, as for certain diseases the risk predictions are more similar among the three companies (e.g. celiac disease) than in others (e.g. prostate cancer) (Kalf et al., 2014). If "genes for" disease existed, things would be

TABLE 8.2 *Average Population Risks and Number of SNPs Used by 23andMe, deCODEme, and Navigenics in the Prediction of Risks for Six Multifactorial Diseases*

Diseases	Average population risk (%)			Number of SNPs used		
	23andMe	deCODEme	Navigenics	23andMe	deCODEme	Navigenics
Macular degeneration	6.5	8	3.1	3	6	6
Atrial fibrillation	27.2	25	26	2	6	2
Celiac disease	0.12	1	0.06	4	8	10
Crohn disease	0.53	0.5	0.58	12	30	27
Prostate cancer	17.8	16	17	12	26	9
Type 2 diabetes	25.7	25	25	11	21	18

Source: Data based on Kalf et al. (2014, p. 87).

simpler, wouldn't they? But they do not, and this is why things are not simple at all.

Results like these should make consumers concerned about the conclusions that can actually be drawn from these tests. They also set the demand for regulations that would protect them, and also ensure that they understand the actual conclusions that can be drawn from them (see e.g. Caulfield & McGuire, 2012). In August 2013, 23andMe launched a national television commercial, which suggested that people could learn a lot about their health and risk for disease for $99 (www .ispot.tv/ad/7qoF/23-and-me). But in November of the same year, the Food and Drug Administration (FDA) ordered 23andMe to stop marketing the respective kit until it received authorization, and the company complied. This happened because 23andMe failed to respond to questions from the FDA about the analytical validity (whether the test can accurately detect whether a genetic variant is present or absent) and the clinical validity (how well a genetic variant is related to the presence, absence, or risk of a specific disease) of that test. FDA requires that companies selling health-related medical devices to the public demonstrate that these are safe and effective. Whereas collecting saliva and

TABLE 8.3 *Agreement among the Three Companies in Assigning Individuals to the Same Risk Category*

	Assigned to the same risk category by all three companies			Assigned to the same risk category by two companies			Assigned to different risk categories
Diseases	↓↓↓	–	↑↑↑	↑↑- ↓↓-	–↑ –↓	↑↑↓ ↓↓↑	↑-↓
Macular degeneration	52.3	0.5	15.2	6.1	6.0	12.5	7.4
Atrial fibrillation	42.4	6.7	16.7	27.3	5.7	1.2	0.0
Celiac disease	75.3	0.0	13.8	9.0	0.4	1.3	0.3
Crohn disease	51.8	0.2	3.5	13.8	3.7	19.9	7.2
Prostate cancer	15.6	4.5	13.5	29.4	21.7	6.5	9.0
Type 2 diabetes	22.2	7.8	14.7	24.1	23.1	3.2	5.0

Note: According to the risk categories used by 23andMe, which categorizes disease risks as decreased (↓), increased (↑), and average (-) risks if the risks of disease are lower than 20 percent below the average population risk, higher than 20 percent above the average population risk, and in between, respectively (values are percentages). For example, ↓↓↓ indicates the percentage of individuals who were at decreased risk according to all three companies, and ↑-↓ indicates the percentage of individuals for which the three companies predicted risks in all three different risk categories.
Source: Data based on Kalf et al. (2014, p. 91).

sending the sample for DNA analysis does no harm to the individual, the interpretation of the results and the handling of the respective data might do.[4] In response to the FDA's order, 23andMe started providing not health-related genetic risk assessments but only genetic data (i.e. genotypes) and ancestry-related genetic reports. The important question is how to find a balance between providing useful information to consumers without imposing barriers to their access to information (Annas & Elias, 2014; Green & Farahany, 2014).

Results like these also raise a series of broader questions: how useful are the results of GWAS for clinical practice? Have perhaps the GWAS been a failure in this respect? The main problem is that GWAS can

---

[4] A very important, relevant issue is that of privacy of genomic data. For an interesting discussion see chapter 9, "Genomic privacy and DNA data banks," in Annas & Elias, 2015.

point to an association between certain DNA variants and a certain disease in several samples from the same population, and even across different populations. This association might be a positive one, i.e. suggesting that people with the disease tend to also have a particular DNA variant and vice versa. However, this does not mean that this variant is the one that actually causes the disease, or at least that it causes the increase in the risk of developing the disease. As shown in Chapters 6 and 7, in most cases the variants identified could explain only a small portion of the observed variation. This is why the results of the GWAS have caused some disappointment, especially if one considers that the total amount of money spent on candidate-gene studies and linkage analyses in the 1990s and 2000s probably exceeds $250 million, without having much to show for it. Nevertheless, it is true that within a few years GWAS have advanced our understanding of genes and pathways involved in common diseases and other complex characters. Evaluation of course depends on the expectations that one had from these studies (see Evans et al., 2011; Visscher et al., 2012). The results of GWAS have also made researchers wonder whether the initial assumption that common variants are responsible for common diseases is adequate. As soon as it was found that these common variants can explain only a small amount of the observed variation, it was also found that numerous rare variants could also account for the unexplained variation. This has made researchers' opinion diverge over whether most of the variance is hidden as numerous rare variants of large effect or as common variants of very small effect. There are various arguments for and against each of these models. However, these may not be mutually exclusive but complementary (Gibson, 2012).

From my own perspective, as someone who is interested in conceptual issues, GWAS have made a major conceptual contribution: they have helped debunk the myth of "genes for" characters and diseases, by showing that single genes or gene variants do not suffice to account for the variation observed therein. This is not of course what they were designed for, yet this is a very valuable conceptual contribution. All these show that there is still a lot to learn about the actual effect of genes on characters and diseases. One might argue that the results in Table 8.3 were to be expected because these are complex diseases anyway. But as shown in the previous chapter, even in the case of monogenic diseases there is a lot of underlying variation at the DNA level.

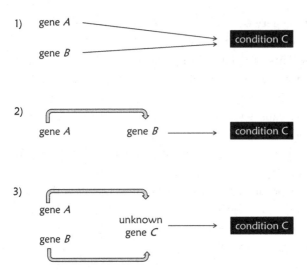

FIGURE 8.1 Some possible biological processes underlying a documented association between genes A, B, and an observed condition C. Until one knows which of the processes (1), (2), (3) is the actual one, there is no real understanding of the effect of genes A and B on condition C.

What is important to understand, however, is that associations between gene variants and disease are rather uninformative insofar as we are unaware of the underlying biological process. Until we are able to elucidate those, all we can find are statistical associations and merely make predictions in probabilistic terms. For example, finding an association between gene A and condition C, and between gene B and condition C, is not really informative if we do not know what exactly A and B contribute to C. Do A and B make independent contributions to C? Or is the contribution of e.g. B dependent on the contribution of A, because A affects B? Or do both A and B affect an unknown gene U, which is actually the major contributor to condition C? Until the precise biological process is revealed, all we can know is the association of A and B with C, without any further understanding. The reason for this is that different biological processes may underlie the same association (Figure 8.1).

Let me further explain why finding statistical associations between genes and conditions is not adequate for understanding. The reason is that a statistical association between two variables may exist without a cause-effect relation between them. For example, the sales of 0.5 litre bottles of cold water and of ice creams rise significantly in Greece during summer. Therefore, one might find a statistical association between the number of ice creams and bottles of cold water sold during summer compared e.g. to the respective sales during winter. Can there be a cause and effect relation there? The answer is yes and no; or better: it depends. It is possible that there is a real cause and effect relation if most people who eat ice cream also need to drink cold water after doing that. In this case, one can assume that the increase in the consumption of ice cream has a causal influence on the increase in the consumption of cold water, because it is the consumption of the former that causes the consumption of the latter. However, it could also be the case that both consumptions increase independently because of the high temperatures (often higher than 35°C) in Greece during summer, and thus because people need to have something chilled to freshen up. In this case, there is no cause and effect relationship between the consumption of ice cream and cold water. The causal factor in this case is the high temperature that makes more people buy both of these. Therefore, in order to clarify what is going on, we need to study in detail the underlying process. Do people who eat ice cream also need to drink cold water afterwards? Or do people buy either of these or both, but without any connection between the two? This is why finding the association alone is not informative if we do not know the underlying causal process.

Therefore, understanding the underlying biological processes and mechanisms is necessary for explaining the emergence of characters or diseases. It is the biological process or mechanism that links the genes and the characters; when we do not understand them, we can only look at the termini: the genotype and the phenotype (Kitcher, 1997, p. 242). But this is not enough. Explaining characters, in turn, requires the identification of the respective causes: what caused a certain character, or a certain version of a character, or a disease to develop. Genes can certainly have a causal role in such cases, but they are not the only causal factors and not necessarily the most important ones. There can be several kinds of causal factors: actions, events, states of affairs, failures to act, background conditions, the nonoccurrence of an event, and the nonoccurrence of a condition. But which of these are selected from the set of factors sufficient to produce an outcome depends on the contexts and on our interests. Let us consider a broken window as an example. If a door with eight glass panes slams and one of the panes breaks, there are several different causal questions that might be asked. If the question asked was why the glass pane broke, then the slamming of the door can be a sufficient causal explanation. But the question asked could be why it was the particular pane that broke and not any of the other seven. In this case, there are several potential explanations. It could be that this glass pane was more fragile because it was made of a different type of glass from the other seven panes; it could be its location on the door that made it more prone to break; or it could be that the framing around that pane was not that good. Finally, we might be interested in why the door slammed. This could happen because someone opened the window in the room and there was wind, or because the people living in the house had a fight and one of them slammed the door. In both cases, the causal inquiry can go further, to a physical explanation of what differences in the temperatures of aerial masses caused the wind, or a psychological explanation of why there was a fight between the people (Cranor, 2013, p. 111). Our world is a multicausal one.

Elucidating the causal role of genes is exactly what the process of "gene-knockout," the targeted disruption of particular genes in their actual biological context, was initially intended to do. Gene knockout involves the replacement of the "normal" copy of a gene with an abnormal copy, which in turn results in the production of a nonfunctional protein or no protein at all. This was expected to produce a phenotype lacking some function that could then be attributed to the knocked-out gene, as its disruption would be the most plausible explanation for the loss of the particular function. However, even in the cases of genes that were considered as essential for some function, the knockout procedure has not always given the anticipated effects, and has thus shown, perhaps in one of the clearest ways, that there are no "genes for" characters. A very likely reason is that many genes are implicated in a certain function, and if one of them is not functioning the others may compensate for it (for various examples, see Morange, 2002). Gene-knockout technology is the combination of two techniques: the culture of multipotent embryonic stem cells from mouse embryos, and the introduction of mutations into these cells by homologous recombination. This is a procedure used in order to insert an allele to a specific homologous genetic locus. It can be performed in embryonic stem cells, which are then injected in an early embryo (blastocyst) and differentiate in all types of cells. As a result, this embryo will give rise to a chimeric organism, i.e. one that has two types of cells: the "normal" ones and those with the knockout gene. This chimeric organism can later breed and give rise to offspring that will have the knockout gene in all their cells (Hall et al., 2009).

One application of this procedure was in immunology with the aim of understanding the function of the major histocompatibility complex (MHC), a set of cell-surface proteins in lymphocytes. Except for their role of presenting antigens to T-cells, it was assumed that MHC class I genes were required for T-cell development. Already in 1990, mice in which the *B2m* (beta-2-microglobulin) gene was disrupted were generated, with no MHC class I molecules produced. The mutant animals developed normally, but had very reduced numbers of cytotoxic T lymphocytes (CTLs). This indicated that MHC class I molecules are not essential for development, but that they are needed for the selection and function of CTLs (Koller et al., 1990; Zijlstra et al., 1990). Then studies of mice deficient for the T-cell co-receptor CD8 showed that it was required for the development of CTLs, but not helper T-cells (Fung-Leung et al., 1991). These and other successes were important in better understanding gene function. However, important problems soon emerged: (1) Many genes encode related molecules and so it is difficult to define the precise function of a gene product due to its overlap with others. In this case, multiple gene knockouts are necessary. (2) Knockouts of the same gene with different strategies can result in different phenotypes. (3) Even when the same knockout strategy is used, differences in the genetic constitution of mice can produce differences in phenotypes. (4) The disruption of a gene might affect the expression of another one close to it, thus producing an altered phenotype that is not connected to the gene of interest. (5) The same mutation in different animal colony conditions can give rise to different degrees of disease occurrence. (6) Disruptions in genes that affect an organ, such as the lymph nodes or the spleen, might have secondary effects on the function or differentiation of its cells, which might be misinterpreted as primary effects of the gene disruption on the cells themselves (Mak et al., 2001). These observations once more indicate clearly that the relation between genes and phenotypes is not a one-to-one but a many-to-many one.

This kind of research with knockout mice has improved our understanding of human disease.[5] Yet, from the results in mice we can only

---

[5] The moral issues surrounding experimentation and genetic engineering with mice and other animals are important but fall outside the scope of this book (see Rollin, 2006, for a discussion of several relevant ethical issues).

make inferences about the respective human conditions, as there are important differences between mice and humans. At the same time, experimentation with humans is, of course, not an option. However, there exist "natural knockouts" among humans, i.e. people with mutations that have resulted in the loss of function for particular genes, the study of whom has yielded some surprising conclusions further indicating the complexity of the relation between genes and characters. Apparently, such mutations have resulted in disease, including those discussed in the previous chapter. However, modern sequencing methods that span the whole genome have also allowed the identification of human knockouts without any accompanying disease phenotype (see Alkuraya, 2015, for a review). An interesting observation has been that even for genes that have a well-established connection to disease, natural human knockouts have very different phenotypes than those of individuals having only one functional allele of the respective gene (this phenomenon is called haploinsufficiency). For example, carrying a certain *BRCA2* allele is associated with a risk for breast cancer. However, human knockouts for *BRCA2* have different phenotypes such as primordial dwarfism, a disease with severely impaired fetal growth that results in shorter size in adulthood (Shaheen et al., 2014). Interestingly, it has been estimated that human genomes contain approximately 100 genuine loss-of-function variants with approximately 20 genes completely inactivated (MacArthur et al., 2012). It seems that this phenomenon is quite common and varies depending on the organ or tissue involved as e.g. genes that are highly expressed in the brain are less often completely knocked out than other genes (see e.g. Sulem et al., 2015). Apparently, there is still a lot to learn.

Insofar as we do not understand all biological processes in detail, all we are left with are associations between genes and characters (or diseases). In order to describe the phenotypic variation in these, new concepts such as penetrance and expressivity were invented. These two concepts can be confused but they are very different. Penetrance is about how many individuals with the same genotype exhibit the same phenotype, whereas expressivity is about the qualitatively different phenotypes that individuals with the same genotype can exhibit (Botstein, 2015, pp. 23–4). On the one hand, penetrance is the proportion of individuals with a given genotype who have a typical associated phenotype. For β-thalassemia, penetrance could be 100 percent as an individual with two defective *HBB* alleles will probably have the disease. However, in other cases such as breast cancer the penetrance of the *BRCA1* alleles is, on average, 50 percent at the age of fifty and 85 percent at the age of seventy. In other words, penetrance is the concept indicating that having an allele related to a disease does not necessarily entail having the disease; it also describes how likely this is. On the other hand, expressivity is about qualitative differences in the phenotypes of individuals with the same genotype, or in other words the degree to which a particular phenotype appears given a certain genotype. For example, some people with a genotype associated with a disease e.g. with the same *PAH* alleles may experience mild symptoms of PKU whereas others may experience severe symptoms. These concepts of penetrance and expressivity were invented in order to accommodate the variability produced by the processes of development. The phenotype of an individual may indicate a genotype, but it is developmental processes that actually produce the phenotype. This is why inferences about the presence or absence of a certain gene from the phenotype may be wrong (see Chapter 9).

Another important phenomenon, also revealed by GWAS, is pleiotropy. In the previous chapter I mentioned genes that may be associated with, and thus assumed to be implicated in, more than one disease (Figure 7.7). More broadly, there exist associations between gene variants and more than one character, and these are described as cross-phenotype associations. However, in these cases we do not know the underlying cause of the observed association. Pleiotropy occurs when a genetic variant *actually* affects more than one character, and it is thus a possible underlying cause for an observed cross-phenotype association. In other words, pleiotropy can be identified only when we have understood the underlying mechanism. Overall, we can distinguish between three different types of pleiotropy. The first is biological pleiotropy that occurs when a genetic variant or gene has a direct biological effect on more than one phenotype. For example, the common coding variant in *PTPN22* (protein tyrosine phosphatase, nonreceptor type 22) on chromosome 1, which has been found to be associated with disorders related to the immune system (such as rheumatoid arthritis, Crohn disease, systemic lupus erythematosus, and type 1 diabetes), seems to influence the function of various types of T-cell and to also be involved in the removal of auto-reactive B-cells. A second type is mediated pleiotropy that occurs when one phenotype is itself causally related to a second phenotype, so that a genetic variant or a gene associated with the first phenotype is indirectly associated with the second phenotype. For example, in the previous chapter I mentioned that several genetic variants have been found to be associated with both LDL levels and the risk of coronary heart disease. However, LDL levels are themselves risk factors for coronary heart disease. Therefore, what must be clarified is whether a genetic variant influences coronary heart disease risk by altering LDL levels or whether it has another effect that is independent of LDL levels. Finally, spurious pleiotropy occurs when a genetic variant falsely appears to be associated with multiple phenotypes, but it is actually not related. This is usually due to faults in the study design such as collecting a nonrandom subsample with a systematic bias so that results based on it are not representative of the entire sample (ascertainment bias) or when individuals with one phenotype are systematically misclassified as having a different phenotype (Solovieff et al., 2013).

More recent findings are even more indicative that the gene-character relation can be very difficult to understand. A study that focused on healthy individuals aimed at identifying whether there could be healthy carriers of gene variants associated with highly penetrant forms of disease. The comprehensive screening of 874 genes in 589,306 genomes led to the identification of 13 adults who had mutations for 8 severe Mendelian conditions, but no reported clinical symptoms of the respective disease (Chen et al., 2016). This clearly shows that having a certain gene or variant does not necessarily bring about the disease. Another recent study analyzed sequence data from 7,855 clinical cardiomyopathy cases and 60,706 ExAC reference samples (see also Chapter 7), in order to better understand the range of genetic variation in this disease. It was found that the variation of certain genes, which were previously reported as important causes of cardiomyopathy, was not clinically informative because of a very high likelihood of false-positive results (Walsh et al., 2016). These findings in turn raise concern about the criteria used to identify pathogenic variants.

This is why particular guidelines have been proposed in order to investigate the effect of DNA variants in human disease in a reliable manner, and thus evaluate the evidence indicating that a variant has a causal effect on a disease. This is very important because research has shown that DNA variants may be related to disease in a variety of ways. In particular, a variant may be: (1) associated with the disease, i.e. be found in significantly higher frequency in patients compared to matched controls; (2) implicated in the emergence of the disease, when there is adequate evidence consistent with a pathogenic role; (3) pathogenic, when it contributes mechanistically to disease, but is not alone sufficient to cause the disease (MacArthur et al., 2014). Therefore, in order to understand the impact of DNA variants on complex phenotypes, more detailed and extensive data is required. This should include detailed clinical descriptions, more information about the associated genotypes, and tissue-specific expression data from large cohorts of patients (Weischenfeldt et al., 2013). It seems that the level at which attention should be paid is that of the cell and not of DNA. This would allow for a more holistic view of the role of DNA as a cell-memory and as part of the control system that uses it. This may be achieved by what is described as "cellular phenotyping," or phenotype at the cellular level. This is the quantification of processes and functions at the cellular level (e.g. gene expression, metabolite production or mitotic rate) and represent, to a large extent, the state of the cell. These processes and functions can then be analyzed like other characters, such as height or cholesterol levels. This would allow for an analysis of the molecular consequences of functional genetic variants and for an assessment of environmental effects at the cellular level, thus shedding light on the biology of complex characters and diseases (Dermitzakis, 2012).

From all these, it becomes clear that current genetic tests can provide some limited information about the probability to develop a character or a disease, and in no way can they predict whether this will happen or not. This is very important to keep in mind as the potential to develop genetic tests may allow at the same time the invention of an unlimited number of diseases. Apparently, genetic tests can be a source of income for the companies that produce and sell them. In addition, physicians might prefer to prescribe a genetic test, although it may not be necessary, in order to refrain from leaving themselves open to charges of negligence. Worse, the possibility of genetic danger might enhance a fatalistic attitude toward disease, instead of motivating people to improve their health behavior (Hubbard & Wald, 1997, p. 27, pp. 66–70). For example, a woman who smokes had better think about quitting smoking rather than be tested for the *BRCA1* and *BRCA2* alleles, as the contribution of the former to the development of cancer is generally more probable. If this woman found out that she does not carry any of these alleles, she might continue smoking; such an irrational attitude had better be avoided.

As Evelyn Fox Keller has pointed out, the question "what genes do?" may actually make no sense. First, genes do nothing on their own; they must be activated or inactivated by particular molecules (mostly proteins) produced by the expression of other genes in order to produce the RNA or protein molecule that in turn is implicated in some process that results in a character or disease. Second, as explained in Chapter 4, genes cannot even be isolated and analyzed; it is DNA that can be isolated and analyzed. Yet, even DNA cannot do anything on its own either, and it certainly cannot directly produce characters. It is more appropriate to think of DNA as the primary resource on which cells draw. But even in that case, DNA is part of a complex system of interacting molecules that all together guide the development of characters (Keller, 2000, p. 57, 2010, pp. 50–51). Therefore, if we would like to summarize the several, often overlapping, roles of DNA within cells, and answer the question "What do genes do?" we might conclude that: (1) genes are templates for the production of functional molecules (proteins or RNA), thus having the role of cell-level memory; and (2) they are part of a control system, the whole cell, which controls the use of this memory and regulates which proteins are produced in it and when this happens (Godfrey-Smith, 2014, pp. 88, 93; see also Morange, 2002, p. 24). These roles are realized through the implication of gene products in developmental processes. But development is not always carefully considered in discussions about genes, character and disease. This is a big omission because if genes do anything at all, it is that they are implicated in the development of characters. It is to this topic that we now turn.

9 Genes Are Implicated in the Development of Characters

Perhaps the most prominent illustration that genes do not determine characters in a straightforward manner is provided by the unfulfilled expectations of the Human Genome Project (HGP). These expectations were enormous and were based on a conception of genes as the absolute determinants of characters. Watson, the first director of the HGP, noted in 1992: "Since we can now produce good genetic maps that allow us to locate culprit chromosomes and then actually find the genes for disease (as Francis Collins found the gene for cystic fibrosis), genetics should be a very high priority on the agenda of NIH research" (Watson, 1992, pp. 167–168). And he concluded: "We have to convince our fellow citizens somehow that there will be more advantages to knowing the human genome than to not knowing it" (p. 173). Walter Gilbert (1932–), another proponent of the HGP and codeveloper of a famous DNA sequencing method, considered genes to determine more than disease: "Three billion bases of sequence can be put on a single compact disk (CD), and one will be able to pull a CD out of one's pocket and say, 'Here is a human being; it's me!' But this will be difficult for humans ... To recognize that we are determined, in a certain sense, by a finite collection of information that is knowable will change our view of ourselves" (p. 96).[1] Twenty-five years later none of these expectations has been fulfilled. Nevertheless, the HGP made a significant, unexpected, contribution: it revealed the complexity of hereditary phenomena, as described in Chapters 6–8, and showed that genetic determinism is a problematic notion as there are no "genes for" characters or disease (Sarkar, 2006).

The HGP was based on the premise that phenomena at the organismal level (e.g. characters) can be explained by reference to phenomena at the molecular level (e.g. genes). This way of thinking is described

---

[1] As it was nicely put: "The CD, so familiar to the audience of a 1990s high-tech society, was recruited to symbolize the merger of molecularisation and digitialisation of biomedical research heralded by the HGP" (Rose & Rose, 2012).

as reductionism, and from its inception the HGP was essentially a reductionist project (Vicedo, 1992). Roughly put, there exist two types of reductionism: methodological and metaphysical (ontological). Methodological reductionism claims that the most fruitful strategy to investigate and understand a biological system is its decomposition into the parts of which it consists. This entails a commitment to the required decomposition approach, but not to the nature of the components of the system. In contrast, metaphysical (ontological) reductionism entails a commitment to the nature of the components of the system, i.e. that these are physical components and therefore that only if there is a change in the respective physical phenomena can a change in the biological phenomena occur (Griffiths & Stotz, 2013, pp. 57–58). Both versions of reductionism share the idea that there is a causal chain that begins from the lower levels of organizations (e.g. genes) and proceeds through the subsequent hierarchical levels to produce an outcome such as a phenotype at a higher level (e.g. organism). In other words, according to this view causation proceeds upward, and so factors at lower levels have a causal influence on factors at higher levels of organization (Figure 9.1). This is the way of thinking that considers genes as the ultimate, and therefore main, causes of phenotypes.

Let us now consider if this is the case, by examining in more detail what happens during development. A major confusion arises from the original sense of the term itself. "Development" literally means unfolding something that already exists preformed somewhere. It is no coincidence that the same term is used for the process of printing on paper the image that already exists in a photographic film (Lewontin, 2000, p. 5). During the eighteenth century there were two major competing theories of development: preformation and epigenesis. Preformation theories suggested that a germ capable of development already possessed a certain structure that somehow preconditioned the adult form. In contrast, the theories of epigenesis suggested that a germ capable of development was unformed (Maienschein, 2012; Müller-Wille & Rheinberger, 2012). Richard Lewontin (1929–), one of the most prominent and most effective critics of genetic determinism, has argued that it is actually the preformation view that has dominated genetics. Of course, nobody has thought that an adult form is preformed in the first cells from which the organism develops. Nevertheless, the idea of a "blueprint" that contains the necessary information for the production of an adult form is

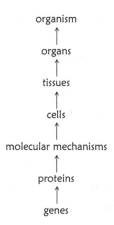

organism
↑
organs
↑
tissues
↑
cells
↑
molecular mechanisms
↑
proteins
↑
genes

FIGURE 9.1 According to the reductionist concept, causation has an upward direction from genes to organismal characters.

quite similar, because it accepts that there exists some fixed genetic essence inside the organism that causally determines its form (as shown in Figure 9.1). Thus, the role of the external environment is limited to certain conditions that may be necessary to trigger the developmental process and to allow it to proceed along a rather predetermined path (Lewontin, 2000, p. 6, pp. 10–13).[2]

However, this is not what happens during development, and this is why the "blueprint" metaphor is actually a very bad one. The development of tissues and organs is not controlled by genes or DNA, but by the exchange of signals among cells. These signals consist of gradients of signaling proteins: "a lot of" protein A might mean "do this"; "some" protein A might mean "do that"; and "no" protein A might mean "do nothing at all." Details notwithstanding, what is important to note is that whatever a cell does and whatever kinds of signals it sends out depends on the kind of signals it receives from its immediate environment. Therefore, neighboring cells are interdependent, and it is local interactions among cells that drive the developmental processes. At the same time, these localized processes also make the development of different organs relatively independent. This allows for control and changes in each organ independently from other organs.

[2] In this sense, genetic essentialism is a necessary condition for genetic determinism, and both of these are necessary conditions for genetic reductionism (see the respective definitions in Prolegomena).

Most interestingly, this is often achieved by using and reusing the same signaling proteins (Davies, 2014, p. 132). To use a simplistic metaphor: the route and the destination of a car is not determined from the place its driver wants to go; it is actually determined from the signals the driver receives. Thus, the driver will stop at a red light; will turn to avoid another car; and may change route in order to avoid traffic. Eventually, the driver may arrive at the intended destination; but the driver may also decide to return home e.g. because of bad weather conditions. This metaphor has problems but it shows that the destination and the route may change because of the signals the driver receives and do not only depend on where he wants to go. In this sense, environmental signals matter for development.

The first and most striking evidence that the local environment matters for the outcome of development was provided by the experiments of Wilhelm Roux (1850–1924) and Hans Driesch (1867–1941). Roux had hypothesized that during the cell divisions of the embryo hereditary particles were unevenly distributed in its cells, thus driving their differentiation. In other words, differentiation during embryo development was due to the differential distribution of hereditary particles in the various cells. This suggested a kind of preformation, and even the first blastomeres would have different hereditary material. As all new cells would be qualitatively different from one another in terms of the hereditary material that they carried, the embryo would become a kind of a mosaic. Roux decided to test this hypothesis. He assumed that if it were true, destroying a blastomere in the two-cell or the four-cell stage would produce a partially deformed embryo. If it were not true, then the destruction of a blastomere would have no effect. With a hot sterilized needle, Roux punctured one of the blastomeres in a two-cell frog embryo that was thus killed. The other blastomere was left to develop. The outcome was a half-developed embryo; the part occupied by the punctured blastomere was highly disorganized and undifferentiated, whereas those cells resulting from the other blastomere were well-developed and partially differentiated (Figure 9.2a) (Allen, 1978a, p. 29; Maienschein, 2014, pp. 70–71). This result stood as confirmation for Roux's hypothesis. As shown in Figure 9.2a, half of the embryo developed normally giving the "left-half," whereas the "right-half" of the embryo did not develop because the respective blastomere was killed. It thus made sense that

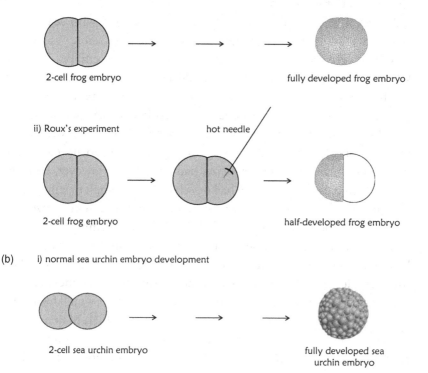

(a) i) normal frog embryo development

2-cell frog embryo   fully developed frog embryo

ii) Roux's experiment   hot needle

2-cell frog embryo   half-developed frog embryo

(b) i) normal sea urchin embryo development

2-cell sea urchin embryo   fully developed sea
                                  urchin embryo

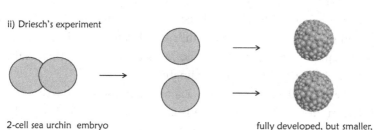

ii) Driesch's experiment

2-cell sea urchin embryo   fully developed, but smaller,
                                  sea urchin embryos

FIGURE 9.2 (a) Roux's experiment: (1) A two-cell frog embryo developing to a normal embryo. (2) Killing, but not removing, one of the two cells resulted in the development of a half-embryo; (b) Driesch's experiment: (1) A two-cell sea urchin embryo develops to a normal urchin embryo. (2) Separating the two cells from each other resulted in the development of each one of them into a complete, but smaller, sea urchin embryo.

the "left-part" and "right-part" hereditary information had already been distributed to the two blastomeres.

However, the experiments that Driesch conducted with sea urchin embryos a few years later pointed to a different conclusion. Whereas it was impossible to separate the blastomeres of a frog embryo without killing it, sea urchin blastomeres could be separated completely just by shaking. So, what Driesch did was to shake a container in which two-cell urchin embryos existed. This resulted in the separation of the blastomeres of each pair, which Driesch let develop on their own. Interestingly, he expected each of the blastomeres to develop to a half-embryo, as Roux's experiments had suggested. However, he soon observed that each blastomere had developed to a complete, albeit a bit smaller, embryo. This showed that each of the blastomeres of a two-cell embryo had the potential to develop to a complete embryo if it were separated from the other one (Figure 9.2b). This could not be possible if hereditary information was distributed differently in each of the blastomeres. However, Driesch observed that if the two blastomeres developed together, each of these would give rise to cells that would form part of a different part of the embryo (Allen, 1978a, pp. 29–31; Maienschein, 2014, pp. 71–73). The important conclusion for our discussion here is that the local environment has an impact on the outcome of development. When a blastomere is on its own, it gives a complete embryo; when it is attached to another blastomere, no matter whether the latter is dead or alive, it gives a half-embryo. This happens because development depends on signals from the local environment that make cells move and differentiate in a coordinated way.

Development comprises two major phenomena: growth and differentiation. In the first case, there is cell proliferation and thus the total number of cells of the body increases, and thus the body grows. Which cells will and which cells will not divide is determined by local signals and interactions. At the same time, cells also differentiate and give rise to the various tissues and organs of the body. Perhaps the most astonishing stage of this process is gastrulation, during which there is a massive reorganization of the embryo from a simple spherical ball of cells, the blastocyst, to a fetus consisting of different types of layers: endoderm, mesoderm, ectoderm. Epithelial tissues and glands of the digestive and respiratory systems are derived from the endoderm; connective tissues (cartilage, bone, blood) and muscle (cardiac, skeletal, smooth)

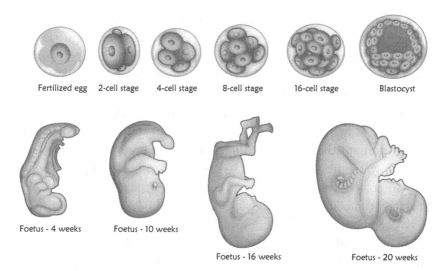

Fertilized egg   2-cell stage   4-cell stage   8-cell stage   16-cell stage   Blastocyst

Foetus - 4 weeks   Foetus - 10 weeks

Foetus - 16 weeks   Foetus - 20 weeks

FIGURE 9.3 Human embryo development. A spectacular change in form takes place in just a few weeks, as an embryo with radial symmetry gradually develops to a fetus with bilateral symmetry (© tofang).

are derived from the mesoderm; and skin and parts of the nervous system are derived from the ectoderm. The process of differentiation is also determined by local signals and interactions. The combination of the processes of growth and differentiation make an early embryo with radial symmetry to develop to a much larger and a lot more complex fetus with bilateral symmetry (Figure 9.3) (Davies, 2014, pp. 38–40).

During development, cells multiply, differentiate, and migrate to various parts of the developing embryo. This happens in a coordinated manner. To understand this, imagine a very large number of humans moving within a limited area in such a way so that no one steps onto the other. How is this possible? There are two plausible explanations. One is that someone organized the crowd and told them how to move in a coordinated way, as would be the case in a military parade. Another is that people simply respond to local cues (see, listen) and avoid stepping on one another, as would be the case while walking in a commercial street on the first day of sales (e.g. walking in Oxford Street on Boxing Day). We now know that the organization of cells in the developing body is achieved in the second way. There is no centralized coordination of development; cells respond to signals from their local environment. Of course, something might go wrong: people might for some reason panic and start stepping and falling onto one another. Under normal conditions though development takes place on the basis of local signals. But then how do genes fit in this picture? Signal production, signal reception, and signal response all depend on proteins, which in turn are produced by the expression of genes. As already mentioned, genes encode proteins that in turn regulate gene expression. Therefore, genes are implicated in this unconscious coordination of development, but they in no way determine its course and its outcomes, which include characters and diseases (see Davies, 2014, pp. 251–252).

An appropriate way to conceptualize the role of genes in development is to think of an organism as an origami (Wolpert, 2011, p. 11). According to this, the DNA of the fertilized ovum is not a blueprint that contains a full description of the adult organism that will emerge from development. Rather, it contains a set of instructions for making the organism, which will affect cell proliferation and differentiation. These instructions are about "how to make" the adult organism, not about "how the organism will look." Therefore, the DNA of the fertilized ovum contains a generative plan, not a descriptive plan.[3] What matters according to this analogy is that what is available are the instructions about when, where, and how to fold the paper in order to make a structure. A description of how the origami structure will look would be entirely useless because it would provide no clues (at least not clear ones) about how to generate it (Figure 9.4). In the same sense, a description of how the adult form will look is useless; what is needed is a set of instructions about how to generate it. Therefore, what happens during human development is not that genes containing the blueprint for the adult human express themselves and thus the organism is constructed to resemble this blueprint. What actually happens is that cells follow the generative plan encoded in genes and the signals they receive from their environment. It is from the combination of numerous local signals

---

[3] In the original analogy, Wolpert (2011, p. 11) actually uses the term "program." However, I consider the term "plan" as more accurate and thus more appropriate. In my view, the term "program" implies instructions and their implementation, whereas the term "plan" is about instructions only. The notion of a genetic program can be very misleading because it implies that, if it were technically feasible, it would be possible to compute an organism by reading the DNA sequence alone (see Lewontin, 2000, pp. 17–18).

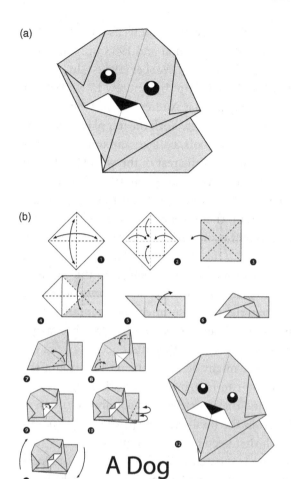

FIGURE 9.4 (a) A descriptive plan; (b) A generative plan. What is encoded in DNA should be conceived as a generative plan, not a descriptive one. (© BlueRingMedia).

coming from the intracellular and the intercellular space that cell division, proliferation, and differentiation take place during development. Thus, from a single fertilized ovum an organism develops. Appropriate signals will drive the production of the anticipated "normal" outcome, whereas "bad" signals can make things go wrong and bring about developmental defects.

Here is another way to conceptualize the role of genes and DNA in development. Imagine a famous play such as, my favorite, *The Phantom of the Opera*. This was originally a novel by Gaston Leroux (1868–1927) published in 1909–1910. However, it became famous as a musical that has been a huge success since 1986 in London and 1988 in New York, based on a play and music by Andrew Lloyd Webber (1948–). In 2004, it was also released as a film, based on a script by Andrew Lloyd Webber and Joel Schumacher (1939–). Imagine that Andrew Lloyd Webber used the same script for the London and the New York theater plays, and that he also sent it to several secondary schools, college, and university theater groups in the United Kingdom and the rest of the world. Although the script (DNA) is exactly the same, the play (phenotype) can be quite different in its various versions. The reason for this is that in each theater group (developmental system) there is a different director and different actors (proteins) who might or might not vary in how the script (DNA) is interpreted (expressed). One group might follow the script exactly as written; another might make minor changes in the words of Christine; another one might change the words and the time of appearance of the Phantom (the copyright license might not actually allow such changes, but let us assume that this is possible). Such changes might result in variations of the same play: the differences between the New York and the London play might be insignificant because of the professionals involved. But you can imagine how creative high school and college students can be, and therefore that it is possible for them to further increase the variation in the final outcome (how happy Andrew Lloyd Webber would be with that need not concern us now).

It is in this sense that what is written in the *Phantom of the Opera* script is not a descriptive plan (e.g. how the play should look) but a generative one (e.g. how many actors are needed, what each one of them has to say and when, etc.). Not surprisingly, the play that the official London, the official New York, a Cambridge secondary school, and an Oxford University theater group would produce based on the same script might differ significantly, as it would depend on who was the director, who read the script, how that person selected the actors and assigned the roles, what instructions that person gave to them and what changes in the original script, if any, that person made. All plays would be the *Phantom of the Opera* in the broader sense, but with differences. What is important is that these differences would be due to differences in the implementation of the script and not in the script itself. In the same sense, the development of an organism should be conceived as the implementation of a plan included in the DNA of the fertilized ovum.

Variations in the outcome of development could exist even if there were no variations in the sequence of DNA, because of the unique interactions between an organism's genome and the external environment(s) in which it lives. Thus, as different plays are possible based on the same script, it is also possible that individuals with exactly the same genotype may have different phenotypes because of environmental influences during development.

The reason for this, and this is the most important point to make, is that it is not genes but cell interactions through signaling molecules that guide development. How a cell responds to a particular signal depends on its internal state, which in turn reflects its developmental history. Therefore, cells respond to the same signal in different ways, and the same signal can be used in various ways in the developing embryo. One of the most striking examples showing how development depends on communication among cells and responses to local signals stemming from a generative plan, rather than on the local implementation of a larger descriptive plan, is the interaction of genes and environment in brain development. The genes that are active in the brain cells during its development do not specify its final structure (e.g. which connections will be made between neural cells and how many of them in each case). Rather, the proteins specified in those genes together make mechanisms that strengthen, weaken, or destroy connections according to the received signals. These mechanisms are constantly refined, thereby changing the way that signals associate with one another. Therefore, the particular neural connections are the outcome of the interaction between certain proteins, encoded by genes, which participate in mechanisms that make neural connections, and their environment. These proteins do not produce predetermined outcomes; if they actually did so, then nothing could be accommodated through learning and experience. Thus, either "bad" genes or "bad" environment can result in mental deficiencies, whereas "good" genes can contribute to a healthy mind only in an environment that provides the stimuli that will lead to the appropriate internal connections during brain development (Davies, 2014, pp. 192–193).

That the phenotype is not an outcome determined by the expression of genes but rather by complex interactions among gene products and cells becomes even clearer in certain syndromes. At some point during development, some cells in a structure that is called the neural crest are further differentiated to form the various structures of the face. But these cells require particular proteins in order to produce new cellular material and multiply, one of which is Treacle. Mutations in the gene *TCOF1* (treacle ribosome biogenesis factor 1) on chromosome 5 result in cell death. Therefore, very few cells exist to form the structures of the face, and the outcome is a condition in which individuals have slanting eyes, underdeveloped cheeks, a small lower jaw, drooping eyelids, and small or absent earlobes (Treacher-Collins syndrome). This syndrome is an example of the complexities that characterize development, and of the connections between genetics, development, and disease. Whereas in popular parlance, stating that there is a gene for the Treacher-Collins syndrome, would make people think that there is a gene causing the respective facial abnormalities, this is not the case. The *TCOF1* gene is not one that is related to face construction but rather to cell proliferation. When Treacle is not operating properly, the neural crest cells do not have the adequate number of ribosomes and eventually become so stressed that they die. Consequently, a smaller number of cells are available for making the face (Dixon et al., 2006; Sakai & Trainor, 2009; Davies, 2014, pp. 103–104).

From all these, it should be clear that what matters is not only whether a gene is expressed, but also in which context the gene product does whatever it is that it does. This context, in turn, can be influenced by the immediate cellular and the broader tissue-organismal environment. Genetic determinism assumes that a gene determines a phenotype, and that the environment in which the cell or the organism lives makes no difference. This can be the case in laboratory populations, such as those that Morgan and his colleagues studied, where the environmental conditions were controlled and were thus uniform. However, this is not at all what happens in natural populations. The representation of the different phenotypes of a particular genotype in different environments is described as a reaction norm. Quite often, such representations emphasize the role of genes, as if different environments made no difference in the developmental outcome (see Lewontin, 2000, pp. 28, 34; Kaplan, 2000, pp. 29–52; Kitcher, 2003, pp. 285–286).

However, the study of natural populations shows that this is not the case. A classic example of how the same genotypes can actually produce different phenotypes in different environments is the experiment with *Achillea millefolium* plants. Seeds from eighty-one natural populations of *Achillea* were collected, germinated, and grown at Stanford. Then thirty individuals from fourteen populations were cloned and planted at three different stations at Stanford (elevation 30 m), Mather (elevation 1,400 m), and Timberline (elevation 3,050 m). For each of these individuals, morphological and physiological characters were recorded over a three-year period. The results of this experiment showed variability in morphological and physiological characters among the three populations in the different elevations (Núñez-Farfán & Schlichting, 2001). These results show that genetically identical individuals (assumed to have identical genotypes for all those DNA sequences related to stem height) do not necessarily have the same level of expression of the same character in all environments (there were different stem heights at different elevations). Therefore, the phenotype is not invariantly determined by genes; rather, it is the outcome of the interaction between genetic and environmental factors (Lewontin, 2000, pp. 20–22).

How is it possible for a given genotype to give such different developmental outcomes in different environments? This is absolutely natural. The problem is that this is not often discussed in genetics education and outreach. Development actually has two distinct and complementary aspects: robustness and plasticity. Developmental robustness is the capacity of individuals of the same species to exhibit the general characteristics of their species irrespective of the environment they live in, thus resulting in the consistency of phenotype in different environments. Developmental plasticity is the capacity of individuals of the same species with the same genotype to exhibit phenotypic variation, and thus to produce different phenotypes during development as a response to local environmental conditions (Bateson & Gluckman, 2011, pp. 4–5, 8). Let me get back to the *Phantom of the Opera* example used earlier. I explained that different theater groups and directors following exactly the same script might end up with a quite different play. All plays would be the *Phantom of the Opera* (this is robustness); however, one of them might emphasize more the character of Christine whereas another might emphasize more the character of the Phantom, thus diverging from Andrew Lloyd Webber's script (this is plasticity). Following the same script does not guarantee producing the same play if different directors and actors are involved. This is why not only the script (DNA), but also how directors and actors will use it (development) matters.

Furthermore, organisms are a lot more complicated than theater plays. They are dynamic systems, which are characterized by molecular interactions of various kinds. The really important biological molecules in these interactions are proteins; it is proteins, not DNA, that actually "do" something. One might of course argue that even if it is proteins that perform functions, these are encoded in DNA. Nevertheless, DNA is "read" only when it is transcribed and translated by proteins. Furthermore, as already explained in Chapter 4, there are several ways in which particular DNA sequences can be transcribed (e.g. alternative splicing) and/or translated (e.g. post-transcriptional modifications such as RNA editing); therefore, there can be multiple alternative readings of the same DNA sequence. What is even more important is that, even if we accepted that genes are important because they "determine" the structure of proteins, we should not forget that the function of proteins depends not only on their own properties but also on the chemical properties of water, lipids, and several other molecules not encoded in genes, as well as of self-assembling complex systems such as membranes. Genes encode information for the amino acid sequence of proteins, not for their final three-dimensional structure that also depends on the local conditions (Noble, 2006, pp. 34–35).

One might raise an objection at this point: it is indeed possible that some characters are plastic and therefore possible to be influenced by the environment. The height of the *Achillea millefolium* plants mentioned could indeed be different in different elevations. But could the type of flowers be affected in the same way? To take an example closer to home: think of our height and the color of our eyes. Everyone would probably agree that environmental factors, such as nutrition, could make a difference in height. Thus, it would be possible, even though it were less likely, that e.g. two identical twins had a different height because their nutrition differed significantly. But could the environment affect the color of their eyes? The answer is yes, even though eye color could be less likely to be affected by the environment than height. This does not mean that one could be born with brown eyes, and end up having blue eyes in adulthood. Rather, it means that one may have the DNA sequences associated with brown color, but be born with a lighter color because for some developmental reason melanocyte formation in the eyes did not go as it would be expected on the basis of DNA sequences alone (see Chapter 6). The important point made here is that genes indicate a potential or a predisposition; however, whether or not this is realized depends on the developmental processes that in turn depend on the environment. This is why the idea of genetic determinism is wrong.

For these reasons, genes should not be considered as our essences. First, as shown in Chapters 7 and 11, genes are not fixed entities that are transferred unchanged across generations. Changes that affect both the information that they encode and their expression can occur from one generation to the next. In addition, genes do not simply specify characters from which their existence can be inferred. For example, when *Arabidopsis thaliana* plants with the same genotype developed under different conditions, some being exposed to mechanical stimulation and other not, they came to have different characters. In particular, those plants stimulated by touch developed shorter petioles and bolts (a developmental response known as thigmomorphogenesis that may be an adaptive response to wind, precipitation, and/or attacks by insects) (Braam & Davis, 1990; Pigliucci, 2005).[4] In this case, just seeing the different characters of these plants might make one infer that they had different genotypes, yet this conclusion would be wrong. Therefore, genetic essentialism is also wrong, as it is not certain that the existence of particular genes will result in particular phenotypes, and thus their existence cannot necessarily be accurately inferred by the observed phenotypes.

What genetics research consistently shows is that biological phenomena should be approached holistically, at various levels. For example, as genes are expressed and produce proteins, and some of these proteins regulate or affect gene expression, there is absolutely no reason to privilege genes over proteins. This is why it is important to consider developmental processes in order to understand how characters and disease arise. Genes cannot be considered alone but only in the broader context (cellular, organismal, environmental) in which they exist. And both characters and disease in fact develop; they are not just produced. Therefore, reductionism, the idea that genes provide the ultimate explanation for characters and disease, is also wrong. In order to understand such phenomena, we need to consider influences at various levels of

---

[4] This is a good illustration of robustness and plasticity. All plants exhibited the same general characters (robustness) but differed in the length of petioles and bolts as a response to environmental conditions (plasticity).

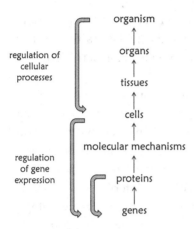

FIGURE 9.5 There are several kinds of interactions among the various levels of organization of organisms. Therefore, we need to consider not only what processes genes affect, but also what affects their own function.

organization, both bottom-up (Figure 9.1) and top-down (Figure 9.5). This is why current research has adopted a systems biology approach (see Noble, 2006; Voit, 2016 for accessible introductions).

All this shows that developmental processes and interactions play a major role in shaping characters. Organisms can respond to changing environments through changes in their development and eventually their phenotypes. Most interestingly, plastic responses of this kind can become stable and inherited by their offspring. Therefore, genes do not predetermine phenotypes; genes are implicated in the development of phenotypes only through their products, which depends on what else is going on within and outside cells (Jablonka, 2013). It is therefore necessary to replace the common representation of gene function presented in Figure 9.6a, which we usually find in the public sphere, with others that consider development, such as the one in Figure 9.6b. Genes do not determine characters, but they are implicated in their development. Genes are resources that provide cells with a generative plan about the development of the organism, and have a major role in this process through their products. This plan is the resource for the production of robust developmental outcomes that are at the same time plastic enough to accommodate changes stemming from environmental signals.

Now, although I argue that genes do not determine characters, and that genes are neither our essences, nor the ultimate explanations for the

(a)

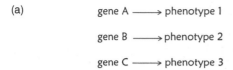

(b)

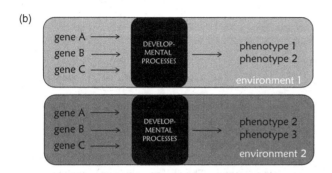

FIGURE 9.6 (a) The common representation of gene function: a single gene determines a single phenotype. It should be clear by what has been presented in the present book so far that this is not accurate. (b) A more accurate representation of gene function that takes development and environment into account. In this case, a phenotype is produced in a particular environment by developmental processes in which genes are implicated. In a different environment the same genes might contribute to the development of a different phenotype. Note the "black box" of development.

development of characters, it is indeed the case that changes in single genes can have a big phenotypic impact. But, whereas a change in a gene may even disrupt a whole developmental process, the final outcome is in no way determined by this gene alone. One interesting case is the *SRY* (sex-determining region Y) gene on the Y chromosome. As already explained in Chapter 2, all humans have twenty-three pairs of chromosomes, and forty-six chromosomes in total. Men and women differ in that women have two X chromosomes, whereas men have an X and a Y chromosome. The default developmental outcome for the human embryo is to become a female. What makes the difference in what will happen is the expression of the *SRY* gene. This gene affects a pathway that guides the development of the male or the female sex (Figure 9.7).[5]

[5] Sex, in the biological sense, should not be conflated with gender. The former is defined on the basis of biological characters, mainly the presence of certain reproductive and sexual organs. In contrast, the latter is defined on the basis of social and cultural characters rather than biological ones. Thus, it is possible for someone to consider oneself as male or female, even if one has ovaries or testes, respectively.

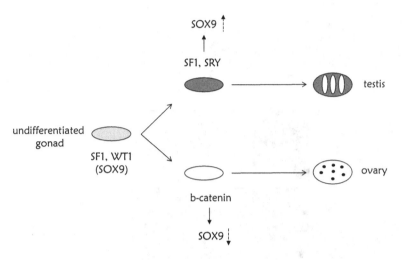

FIGURE 9.7 Sex differentiation in mammals. The SRY protein is only one of the proteins involved in this process.

In this sense, *SRY* makes a difference for the development of sex. Embryos carrying the Y chromosome and the *SRY* gene develop testes and a male reproductive system, whereas those not having either the Y chromosome or the *SRY* gene develop ovaries and a female reproductive system (Davies, 2014, pp. 147–151). Interestingly, it has been found that a mutation (four-nucleotide deletion) in the *SRY* gene was adequate to make an XY individual develop as a female with underdeveloped reproductive organs (Jäger et al., 1990), as well as that a translocation of part of the Y chromosome including the *SRY* gene onto the X chromosome was adequate to make an XX individual develop as hermaphrodite (carrying both male and female organs) (Margarit et al., 2000).

However, the impact of the *SRY* gene seems easy to exaggerate. Consider the following statements:

> In genetic terms, this suggests a peculiar paradox. Sex, one of the most complex of human traits, is unlikely to be encoded by multiple genes. Rather, a single gene, buried rather precariously in the Y chromosome, must be the master regulator of maleness. (Mukherjee, 2016, pp. 359–360)
> Is all of sex just one gene, then? Almost. (Mukherjee, 2016, p. 362)

These quotations about the *SRY* gene from a best-selling book on the gene that aims at explaining the concept to a broad audience are characterized by a conceptual flaw: the nondistinction between genes as character makers and genes as character-difference makers, which was discussed in Chapter 2 and that was clear to Morgan and his collaborators more than 100 years ago. As shown in Figure 9.7, several proteins (and therefore genes) are involved in the process of sex differentiation. The bi-potential precursor of gonads (testes and ovaries) is established by proteins including SF1 and WT1, the early expression of which might also initiate that of SOX9 in both sexes; b-catenin can begin to accumulate at this stage, and in XX cells its levels could repress SOX9 production. However, in XY cells, increasing levels of SF1 activate the production of SRY that with SF1 enhances SOX9 expression. If SRY activity is weak, low, or late, there is no SOX9 expression as b-catenin levels accumulate to shut it down. In the testis, SOX9 promotes the testis pathway, and it can do so even in the absence of SRY (Sekido & Lovell-Badge, 2009). Therefore, the *SRY* gene does nothing on its own and it is not all-powerful. However, it is indeed the case that mutations that involve this gene can have an enormous impact on the development of an individual. It is therefore changes in genes that can have an impact on developmental processes, and genes can act as difference makers that affect the final outcome of development (this is discussed in detail in Chapter 10).

The take-home message should be obvious: genetics, and genetic inheritance, makes more sense in the light of development. Or, most distortions and misunderstandings of how genetic inheritance takes place have occurred because of the failure to take developmental processes into account. Genes are implicated in the development of characters and disease, but do not determine any of them in any way. Developmental processes show that neither genetic determinism nor genetic essentialism nor genetic reductionism is accurate. It is therefore important to think of genes not as the ultimate determinants of who and what we are, but as the resources on which our cells draw and thus develop and live under particular environmental conditions. Having one or the other gene or DNA variant could be informative for one's life, health, and identity. But it could also be meaningless. What is certain is that to know ourselves we need to know a lot more than our genes.

To return to the HGP with which this chapter began: did it go wrong, and where? It seems that the HGP changed the way biological research is done in fundamental ways. It also resulted in new sequencing technologies and their outstandingly fast-paced evolution. Finally, it produced a better understanding of the relation between genes and characters, and highlighted the importance of developmental processes. The progress and the changes in these aspects have been enormous. But as far as medical practice is concerned, there is a long way ahead. This has caused some pessimism, but the outcomes have shown that there is also great potential (see Evans et al., 2011; Lander, 2011). This is why it is important to make the distinction between achievement and wishful thinking, between where we are and where we would like to go. A first step is to realize that the contribution of genes makes more sense in the light of development, as what genes "do" (if they do anything) is only to be implicated in developmental processes through their products. The next step is understanding what genes can really account for, which is the topic of the next chapter.

10 Genes Account for Variation in Characters

Francis Galton was the first to discuss the contrast between nature and nurture in the context of heredity. Nature referred to the influence of one's biological makeup, whereas nurture referred to the influence of one's upbringing. This distinction became widely circulated when Francis Galton published his book *English Men of Science: Their Nature and Nurture* (1874). He wrote:

> The phrase "nature and nurture" is a convenient jingle of words, for it separates under two distinct heads the innumerable elements of which personality is composed. Nature is all that a man brings with himself into the world; nurture is every influence from without that affects him after his birth. The distinction is clear: the one produces the infant such as it actually is, including its latent faculties of growth of body and mind; the other affords the environment amid which the growth takes place, by which natural tendencies may be strengthened or thwarted, or wholly new ones implanted ... When nature and nurture compete for supremacy on equal terms in the sense to be explained, the former proves the stronger. It is needless to insist that neither is self-sufficient; the highest natural endowments may be starved by defective nurture, while no carefulness of nurture can overcome the evil tendencies of an intrinsically bad physique, weak brain, or brutal disposition. Differences of nurture stamp unmistakable marks on the disposition of the soldier, clergyman, or scholar, but are wholly insufficient to efface the deeper marks of individual character. (Galton, 1874, pp. 12–14)

Galton thus conceptualized nature and nurture as being associated with prenatal and postnatal influences, respectively. But these were not two distinct kinds of influences, one responsible for the transmission of characters (nature) and the other for their development (nurture). Rather, he thought of them as two different sets of influences, both of which were necessary, but of which nature was always stronger. In principle, it would be very difficult to isolate and identify the influence of either nature or nurture, as individuals inherit from both their parents. For instance, was the son of a renowned musician a talented musician as well because of biologically inherited factors or because of the musical environment in which he was brought up? As is obvious from the previous excerpt, Galton tended to accept the former. However, he conceived of an appropriate way to establish this: the study of "identical" twins. He thought that as these had the same nature but different nurture, they might provide the clues for distinguishing between the influence of each. Galton thus sent questionnaires to many parents of twins, aiming at establishing how their life experiences affected the similarities and differences between twins (Burbridge, 2001).

It is important to note that "identical" twins are not really identical, genetically speaking, as they do not really have 100 percent the same DNA for two reasons. The first is that, as shown in Chapter 7, a certain number of new somatic mutations occur in each of us. Therefore, it is possible that after the clusters of cells that will give rise to the twins are separated, certain mutations occur in one of them and not in the other. The resulting difference will be minor, but it exists. The second reason is that epigenetic changes, i.e. changes in DNA and chromosomes other than changes in the DNA sequence itself, take place. These phenomena are explained in Chapter 11. What matters for our discussion here is that the genomes of individuals undergo molecular (genetic and epigenetic) changes during development, and thus are not static. As it has been said, we are all born with a *developing genome* (Moore, 2015, p. 15; more about this in Chapter 11). Therefore, the adjective "identical" initially used by Galton and others is not accurate as a description. It is more appropriate to describe these twins as monozygotic twins, i.e. twins emerging from the same zygote (fertilized ovum). In this way, we can distinguish them from the dizygotic twins, i.e. those emerging from two different zygotes that in turn emerged because two ova were fertilized independently. Dizygotic twins are two entirely different siblings who happened to be conceived at the same time because their mother's ovary released two ova simultaneously. Monozygotic twins emerge from the same fertilized ovum.

Generally speaking, there are three possible ways for monozygotic twins to emerge, by "mistakes" in the early cell divisions (see Figure 9.3). In the first case, it is possible that, after the first divisions of the early embryo (cleavage) that produce a sixteen-cell embryo, some of its cells are separated from the others. In this way, two distinct clusters of cells can be formed, which can each develop to a complete human embryo with its own placenta and membranes. In the second case, after a blastocyst has formed, it is possible for the inner cell mass to be divided into two clusters of cells that will each develop to a complete human embryo. The difference in this case is that these embryos will grow on the same trophectoderm (the external cell layer of the blastocyst). However, each embryo will make its own yolk sac and amniotic cavity, and will thus be safely separated. This is the most common way that monozygotic twins come to be. There is a third case in which monozygotic twins emerge due to differences in the patterning of the early bilateral embryo, but it is very rare (see Davies, 2014, pp. 27, 30, 43). In all three cases, monozygotic twins initially have more or less the same DNA because they are the outcome of development of cells that are derived from the same zygote.

Figure 10.1 presents the amount of the common DNA within a family. It must be noted that the percentages in Figure 10.1 are averages, with the exception of the parent-offspring comparison because we have all inherited half of our DNA from each of our parents. This means that two siblings share 50 percent of their DNA on average, because there could be extreme cases – at least theoretically. For instance, two brothers could have inherited exactly the same twenty-three chromosomes from their father (e.g. their paternal grandfather's chromosomes), as well as exactly the same twelve chromosomes and eleven different chromosomes from their mother (e.g. one brother got all chromosomes from his maternal grandfather whereas the other got half of these and half of the chromosomes from his maternal grandmother). These brothers would thus have 75 percent of their DNA being the same (Figure 10.2a). To give another extreme example, two sisters could have inherited totally different chromosomes from each of their parents, with the exception of their father's X chromosome. In such a case, one of the sisters could have inherited all her grand-maternal chromosomes through her parents, whereas the other could have inherited all her grand-paternal chromosomes through her parents. In this case, the two sisters would have very different DNA, with only their father's X chromosome being common between them (Figure 10.2b).

The study of twins became the standard method for understanding the contributions of nature and nurture. Of particular interest were studies

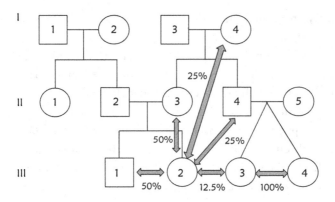

FIGURE 10.1 Relatives have common DNA. The pedigree illustrates the members of a family (males are represented with rectangles and females with circles; spouses are directly connected with a horizontal line; siblings are connected with a horizontal line on top of them that also connects them to their parents). The extreme case for two siblings is to receive exactly the same twenty-three chromosomes from each of their parents, which is very unlikely. There are several possible combinations. As they each receive half of their DNA from each of their parents, which in turn corresponds to half of their total DNA, we can estimate that siblings share on average 50 percent of their DNA (e.g. III1 and III2). Only III3 and III4 share (almost) 100 percent of their DNA because they are monozygotic twins. For example, III2 has also 50 percent of her DNA common with her parents (II3 is her mother), and on average 25 percent of her DNA in common with her maternal grandmother (I4) and her maternal uncle (II4). Finally, III2 and III3 are first cousins and 12.5 percent on average of their DNA is common.
Note: all percentages are averages, except for the parent offspring relations.

of behavioral characters in monozygotic twins. Another approach was to study people adopted as children and compare them to their adoptive and biological parents. A topic that brought about a major controversy was the study of IQ (intelligence quotient) in the 1960s and the 1970s. The concept was invented by psychologist William Stern (1871–1938) in 1912, and it was expressed as the ratio of a child's mental age to the child's chronological age, times 100. If the ratio was 1, then a child's IQ would be 100; if the ratio was 1.2, then the IQ would be 120; and so on (Kevles, 1995, p. 79). This practically meant that a child with an IQ higher than 100 would be more advanced and have a mental age that was ahead of this child's chronological age. In the opposite situation, the child would be considered as somehow having a delayed mental development.

In 1969, educational psychologist Arthur Jensen (1923–2012) pub-lished a paper that suggested that differences in IQ among people were

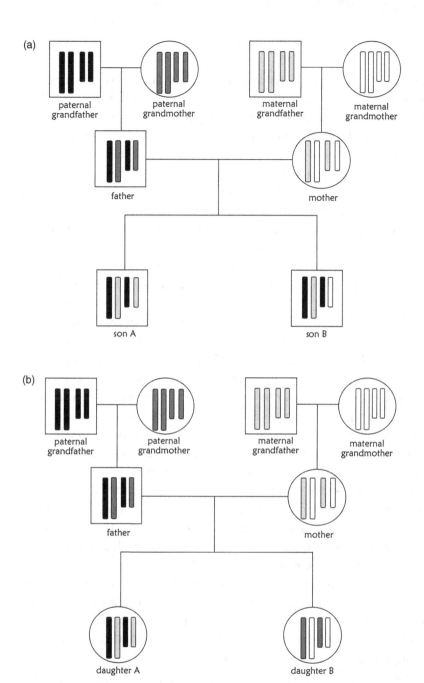

FIGURE 10.2 (a) It is possible for two brothers to share 75 percent of their DNA if they have inherited exactly the same paternal chromosomes and approximately half the same maternal ones. (b) It is possible for two sisters to have entirely different chromosomes if they have inherited one the chromosomes of their grandmothers and the other the chromosomes of their grandfathers, with the exception of their father's X chromosome that would be common to both of them.

largely due to genetic differences (Jensen, 1969). His aim was to question the assumption that IQ differences were almost entirely due to differences in the environment in which people were brought up. By examining the available published data, Jensen concluded that variation in intelligence was mostly due to genetic differences. Jensen compared two extreme cases: monozygotic twins brought up in different families, and unrelated children brought up in the same family. In the first case, the available data showed that the correlation between monozygotic twins in different families was 0.75. Jensen argued that as there should be no genetic differences between monozygotic twins, any difference between them should therefore be due to environmental factors. So, Jensen concluded that 75 percent of the variance could be said to be due to genetic variation,[1] and 25 percent due to environmental variation. He then looked at data from unrelated children living in the same family. Jensen explained that as these children were not genetically related, any correlation between them should reflect their common environment. The available data showed that this correlation was 0.24. Therefore, the proportion of IQ variance due to environment was 24 percent and the remaining 76 percent should be due to heredity. Jensen noted the good agreement between these two estimates: 75 percent and 76 percent of differences should be due to genetic factors in these two extreme cases. He also calculated a composite value from all kinship relations available (not just the two extreme cases discussed), and he concluded, after appropriate statistical corrections, that the best single overall estimate for the heritability of intelligence was 0.81. He thus concluded that approximately 80 percent of the variation in intelligence was due to genetic differences. A major conclusion he made was that the influence of the environment on IQ was that of a threshold variable: extreme environmental deprivation could keep a child from expressing his/her genetic potential, but no enriched educational program could push a child above that potential. Jensen also concluded that educational attempts to boost IQ had thus been misdirected (Jensen, 1969, pp. 50–51, 59–60).

Perhaps the most influential critic of Jensen's conclusions was Richard Lewontin (see also Gould, 1996). In his critique (Lewontin,

---

[1] This was for Jensen the definition of heritability. He wrote: "Thus 75 percent of the variance can be said to be due to genetic variation (this is the heritability)."

1970a), he initially accused Jensen as being biased and as rushing to pre-reached conclusions rather than questioning the methods with which he reached them. Then he argued that it is not possible to speak of characters as being molded by heredity, as opposed to environment. The reason for this, according to Lewontin, is that every character is the outcome of "a unique interaction" between genes and the sequence of environments in which an individual has lived and developed. Sometimes it is genes or environment that are more influential in shaping the developmental outcome, but in all cases the relationship between genes and characters, as well as environments and characters, is a many-to-many one (p. 5). Lewontin went on to explain why we cannot talk of heritability of a character in general, but only in a particular population and in a particular set of environments (p. 6), a point to which I return below. Lewontin concluded his critique by making two major points: (1) that in order to explain how genes and environment affect characters it is necessary to understand the respective developmental process, and (2) that conclusions made about the possible contributions of environment may be wrong because of the limited number of different environments studied and because not all possible environments have been considered (see also Jensen, 1970; Lewontin, 1970b). In subsequent work, Lewontin drew on data such as those on *Achillea* plants, to argue that phenotypes are influenced by differences in both environment and genotype, and that their various combinations produce a variety of phenotypes. In this sense, both genotypes and environment should be considered as causes of phenotypic differences (Lewontin, 1974; Levins & Lewontin, 1985, pp. 109–122; see also chapter 4 of Kaplan, 2000, for a broader discussion).

What is at stake here? If one reads the exchange between Jensen and Lewontin, one may conclude that there are ideological biases behind the different views and the criticisms. In a thoughtful analysis of this exchange, however, Tabery (2014) has argued that the real problem is what he describes as the explanatory divide (p. 65). According to Tabery, the debate was about three distinct questions: (1) What is the interaction between nature and nurture? (the conceptual question); (2) Why and how should the interaction between nature and nurture be investigated? (the investigative question); and (3) What is the empirical evidence for the interaction between nature and nurture? (the evidential question). Jensen and Lewontin addressed these questions differently and this was the real reason of their disagreement according to

Tabery. Details notwithstanding, Jensen was interested in explaining the variation in a population by asking how much of this variation is due to one or the other factor and thus by identifying the causes of this variation. In contrast, Lewontin was interested in explaining the developmental process by asking how this takes place and thus by finding the causal mechanism responsible for it (Tabery, 2014, p. 37, pp. 67–70). Therefore, one important issue has been that the different explanatory aims of scientists may have made understanding one another difficult. However, whereas Lewontin's answers to these questions are very relevant to our discussion here, Jensen's answers are less relevant and may have produced confusion about the concept of heritability.

Consider the following definition of heritability: "The heritability of a character is defined as the proportion of the variance in a population attributable to genetic variation, with the balance presumed to arise from environmental effects" (Slack, 2014, pp. 73). Thus, if we say that the heritability of a character in a population is, say, 0.8 this means that 80 percent of the phenotypic variation in this population is due to variation in genes. This means that this population is genetically heterogeneous, and so 80 percent of the observed phenotypic variation can be explained on the basis of the variety in genotypes, whereas 20 percent of the observed phenotypic variation can be explained on the basis of the variety in environmental factors. A first point to note is that what heritability is about is the explanation of the variation in a character, and not about the origin of the character itself. However, it is easy to conflate these.

To better understand what heritability is about, let us draw on Lewontin (1970a) again, and on two thought experiments he proposed for this purpose. In the first experiment, imagine that we have two populations of corn, say P1 and P2. Owing to inbreeding by self-fertilization, each of these populations consists of individuals with exactly the same genotype (P1 consists only of individuals with genotype *HH*, and P2 consists only of individuals with genotype *hh*). Therefore, within each population there is no genetic variation; however, P1 and P2 are genetically distinct from each other. If we plant seeds in flowerpots with ordinary soil, these will germinate and grow. If after a few weeks we measure the height of each plant, we will see that there is some variation in height among the various plants of the same population. Because the seeds of each population were genetically identical, any difference in height among the resulting plants of the same population must be entirely due to variation in the environmental conditions among the various pots. In other words, the difference in height among plants with genotype *HH* or among plants with genotype *hh* can only be due to the difference in environmental conditions in their pots. For this reason, the heritability of plant height in each population is 0. But as the P1 and the P2 plants have different genotypes (*HH* and *hh*, respectively) and developed under the same conditions, the average difference in plant height between P1 and P2 is entirely due to their different genotypes. Therefore, the difference in height observed between P1 and P2 is entirely due to their genotypes, although the heritability of height in each of these populations is 0 (Figure 10.3a).

Lewontin then suggested the opposite experiment. Imagine that we have two populations of corn emerging from seeds of an open-pollinated variety, which therefore has a lot of genetic variation, i.e. several different genotypes. But instead of using ordinary soil, we grow seeds in vermiculite and we water them with a carefully prepared nutrient used for controlled growth experiments. In this way, there will be genetic variation within each population of plants that will emerge but no environmental variation during their development. Imagine now that we divide the seeds into two groups, and whereas one group develops under the aforementioned conditions producing plant population P3, the other gets half of some essential nutrients and develops producing plant population P4. After several weeks we measure the height of the plants in these populations. The variation in height within each population is entirely due to their genotypes as the environment of the individuals of each population was controlled. Therefore, the heritability for height in each of these populations will be 1. However, the plants of population P3 will on average be higher than the plants of population P4, because the latter did not develop as much as they could because they received a lower quantity of essential nutrients. This means that the difference in height between P3 and P4 individuals will be entirely due to environmental factors, although the heritability of height in each of these populations is 1 (Figure 10.3b). It should be clear that heritability depends on the particular population we are talking about.

Let us consider another example. Imagine a population of cattle in which the heritability of final size is 50 percent. This means that variation in final sizes is 50 percent due to variation in genes and 50 percent

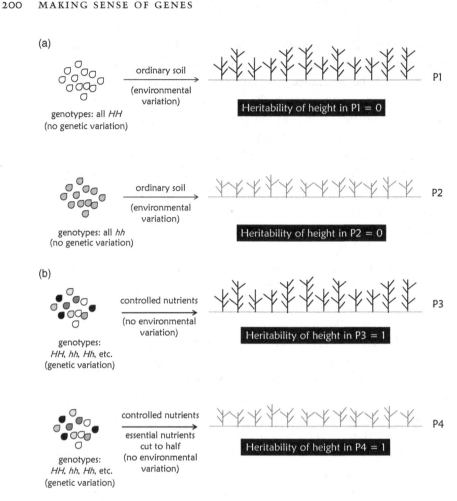

FIGURE 10.3 Heritability: (a) The differences in height between populations P1 and P2 are due to genetic differences, although the heritability of height in each population is 0; (b) The differences in height between populations P3 and P4 are due to environmental differences, although the heritability of height in each population is 1.

due to variation in environmental factors, both of which are implicated in development. Now imagine that at some point a farmer buys all the farms in the area. If the farmer started artificially selecting the same type of cattle, with a very specific final size, after several generations genetic variation would probably be smaller compared to the beginning, because some variants that were not selected would no longer be present in the population. If environmental variation had been kept constant in the meantime (i.e. the differences in nutrition and husbandry among the farms continued to exist during that time), eventually the heritability of final size would fall because of the reduction in genetic variation. In contrast, if the farmer instituted the same differences in nutrition and husbandry in all farms, and thus the environmental variation was reduced, the heritability of the final size would increase. This would be possible, despite no changes in the genetic variation, exactly because the environmental variation was reduced. What this example again shows is that heritability is context-dependent and any measurement is valid only for a particular population under particular conditions (Slack, 2014, p. 75). This and the previous example should make clear that the heritability of a character depends on the population and on the character under consideration. We cannot predict the heritability of a character, even when we know that it is genetic, and we cannot predict if a character is genetic, even when it is known to have high heritability (Keller, 2010, p. 62).

Therefore, heritability is not about measuring the relative, additive contribution of genes and environment to phenotypes. This is impossible to measure because these contributions are interdependent, not independent. Here is a way to illustrate this. Imagine two men who lay bricks to build a wall. In this case, it is quite easy to measure the contribution of each one of them to the final outcome. If man A laid 40 bricks and man B laid 60 bricks to build the wall, we can say that the building of the wall was 40 percent due to man A and 60 percent due to man B. In this case, the causal influence of the two men is independent and thus measurable. However, if man A mixes the mortar and man B lays the bricks, it is not possible to measure their relative contributions. Even if we could measure the quantity of mortar and bricks in the wall, this would not be a measure of each man's contribution as both mortar and bricks are needed for building the wall and cannot be considered separately. In this case, it is impossible to distinguish between the causal influences of the two men, because they interact and depend on each other to build the wall (Lewontin, 1974). In the same sense, we can measure the relative contributions of two children filling a bucket with water, only if each one of them is filling the bucket independently because that is how we can measure how much water each one of them added (e.g. 20 liters and 30 liters, respectively). But if one child is holding the hose over the bucket and the other turns the tap on, it is not possible to measure how much water is due to each one of them as both contributions are necessary in order to fill the bucket and cannot be considered separately (Keller, 2010, pp. 8–9). In this sense, the question about the relative contribution of genes and environment to the production of a character makes no sense, because they do not make independently measurable contributions.

Therefore, the question that makes sense to ask is not about the relative contribution of genes and environment to a character but to character differences in a population. Let us consider another example to illustrate this. Imagine a drummer playing his drums. It makes no sense to ask whether the drumming sound we hear is mostly due to the drummer's competence or to the quality of the instrument he is playing. The reason for this is that the outcome depends on both. However, what makes sense to compare are two different drumming sounds. What is then interesting to explain are not the drumming sounds themselves but the differences between them. In the case of two drummers playing the same instrument, any difference between the drumming sounds would be almost 100 percent due to the drummers. In the case of the same drummer playing different instruments, any difference in the drumming sound would be almost 100 percent due to the instrument. In both of these cases, there is a single difference-maker, the drummer and the drums, respectively. Now, the situation becomes more complicated in the case of two different drummers playing two different instruments. In this case, the difference will be somewhere in between the two previous cases, depending on the competence of the drummers and on the quality of the instruments (Keller, 2010, pp. 34–36). The crucial point here is that what can be explained on the basis of single genes is not a certain outcome but differences between distinct outcomes. In this sense, what we can try to explain are differences in characters (e.g. why one has blue and another has brown eyes), rather than characters themselves (e.g. why one has blue eyes or why another has brown eyes).

As already discussed in Chapter 2, it was clear to Morgan and his collaborators that several genes (at the time still called factors) were necessary to produce a character, but also that a change in one of them was sufficient to bring about a change in the character: "It is this one different factor that we regard as the 'unit factor' for this particular effect, but obviously it is only one of the 25 unit factors that are producing this effect" (Morgan et al., 1915, p. 209). This means that although single-gene variation may explain character differences in a population, the development of a character is not sufficiently explained by the activity of a single gene. In other words, whereas changes in a single gene may cause variation in the respective character in a population, it is not the gene alone that causes the development of that character. In this sense, genes are difference-makers for their effects. What single genes can account for are not the effects themselves, but only the differences among them. Thus, *SRY* is a difference-maker for the development of sex, but not the gene that determines sex (see Chapter 9). This means that a change in the presence of *SRY* (depending on whether a human has a Y chromosome or not), or in its expression (depending on whether the *SRY* gene is expressed or not due to a mutation) can make a difference in the development of testes or ovaries. But in no way is *SRY* the gene that "determines" the development of these organs and sex (as mentioned in the previous chapter, a mutation that would affect SOX9 would impact the development of these organs too).

But what exactly is a difference-maker? Generally speaking, there can be several causes of a particular effect. To take a classic example, a forest fire can be the effect of several different causes such as a lighted match, oxygen, and flammable materials such as dried grass. There will be no forest fire if there is no oxygen because it is necessary for combustion; flammable material such as dried grass is also a cause, because wet or humid material is difficult to burn. However, in the case of an actual destruction of a forest because of a fire, it is a lighted match that is considered to have caused the fire. Nobody usually considers the presence of oxygen or the dried grass as the main causes of the fire, although there is no question that these are causes too. The reason for this is that oxygen and dried grass exist in a forest all the time (at least during summer), but the initiation of a fire is not due to any change in them. In contrast, it is a change in the condition of a match that makes the difference and initiates the fire. Therefore, in the case of a forest fire, there are at least three causes (match, oxygen, dried grass), but it is the lighted match that is the cause that makes the difference – the difference-maker. More generally, what this example suggests is that whereas there can be several different causes for a particular effect, there is only one or a few that are more important than the others – causally speaking – because it is their differentiation that brings about the effect.

In a thoughtful analysis, Waters (2007) has suggested that these causes are more important because they can account for the actual difference, in the case of biological characters for the actual variation in a population. Waters conceived of causes as difference-makers and distinguished between two different kinds:

- **A potential difference-maker**: a variable C that if it were actually differentiated would bring about an actual difference in an effect variable E.
- **The (an, in case of two or more) actual difference-maker**: a variable C an actual variation in which would bring about an actual difference in an effect variable E (with no actual difference in any other variable that could account for the actual difference in E).

In the example of the forest fire, the lighted match is the actual difference-maker.

So, in order to explain an effect, one first needs to identify the potential difference-makers, and then the actual difference-maker among them, i.e. the cause that explains why some variable varies in a population, for example, the lighted match in the forest fire example.[2] It should be noted that the concept of difference-makers applies only to causes in actual populations. Therefore, it makes no sense to identify something as the actual difference-maker without previously identifying either a population with members that actually differ with respect to some variable that is the effect of these difference-makers, or a population with members that differ in terms of such an effect variable at different actual times. It should also be noted that the actual effect is not a single property in a single individual, but a difference of a property within a population. Therefore, for example, in the population of Greece that consists of people exhibiting symptoms of β-thalassemia and people without any such symptoms, the potential difference-makers are the various *HBB alleles* and the actual difference-makers those that were actually found in the particular β-thalassemia patients. This should generally be described as that differences in *HBB* cause differences in the β-thalassemia phenotype (between healthy people and patients), and more generally that differences in X cause differences in Y. The statement "differences in X cause differences in Y" and "X causes Y" are

---

[2] Griffiths and Stotz (2013, pp. 79–84) describe a philosophical disagreement on whether it is only DNA sequences or other factors too that can be difference-makers. The details of this discussion fall outside the scope of the present book. However, my view is that genes are not the only difference-makers for the development of characters.

very different. The latter should be interpreted as the former; but by answering the former one does not also answer the latter (Keller, 2010, pp. 38–9).

Genes are not the only difference-makers, even for diseases that are considered to be genetic. For example, lactose intolerance in humans can be either viewed as a genetic disease or as an environmental disease depending on the population considered. The reason for this is that individuals in different populations could be healthy for different reasons. Some people might lack the allele related to lactose intolerance and consume lactose, whereas others might have that allele but not consume lactose. As a result, lactose intolerance is considered to be a genetic disease in populations in which ingestion of milk products is common and lactase deficiency is rare, because it is the latter that actually makes the difference in the occurrence of the disease (most people do not have the respective allele, and so they have no problem even though they consume milk products). In contrast, lactose intolerance is considered to be an environmental disease in populations in which ingestion of milk products is rare and lactase deficiency is common, because it is the former that makes the difference in the occurrence of the disease (most people do not consume milk products so they have no problem even though they have the respective allele). Therefore, a character cannot be described as "genetic" in any absolute sense, as this depends on the respective population (see Gannett, 1999). This is important because either the presence of an allele or the presence of an environmental factor could make the difference in the occurrence of a character or a disease. What matters is not whether a gene is associated with a character or a disease, but the detailed understanding of why and how changes in a gene might bring about a certain version of a character or a disease (which is defined here as a difference from the perceived "normal" state).

I hope that this clarifies what heritability is and is not (for a more detailed and technical discussion, see Visscher et al., 2008). The following points should by now be clear. First, heritability is about a character in a certain population, not about this character in general. This means that we cannot talk about the heritability of corn height in general, but only about the heritability of height in corn populations P1, P2, P3, P4. Second, even strongly inherited characters can have heritability equal to zero in particular populations. For example, humans normally have two legs and two arms. Differences in DNA that cause differences in the number of limbs in humans are rare, and so most differences are explained in terms of differences in the environment (e.g. accidents leading to amputation, environmental influences during development, etc.). As a result, whereas our DNA indicates that we should have two legs and two arms, instead of one or three, these characters have a very low heritability. These mean, in a nutshell, that heritability estimates do not reflect the importance of genes or environment in the production of one's characters; in fact, heritability estimates tell us nothing about what causes one's characters (see Moore, 2013).

Given this, it is important to avoid the problems caused by the slippage in use and the public understanding of the terms "heritable" and "heritability." In many cases, heritability has come to denote the property of being inheritable. However, as explained, the technical definition of heritability refers to a statistical rather than to a causal measure; it makes no sense to use it for individuals, but only for populations (Keller, 2010, pp. 5–58). As Moore (2002, p. 45) notes, it should be made explicit that heritability is entirely different from inheritability, i.e. how inheritable a character is. Therefore, although having two legs or two arms is an inheritable character, their heritability is close to zero because as explained, differences in the number of arms and legs are more often than not due to environmental influences. Thinking in terms of difference-makers, whereas several genes causally influence the development of legs and arms, it is differences in the environment and not differences in those genes that cause the variation in the number of arms and legs.

The important conclusion of this chapter is that developmental conditions and influences affect why, when, and how our genome is expressed. In other words, our "nurture" affects our "nature"; this does not make the latter less important than the former. Actually, it is only accurate to say that the two interact. Thus, genes do not act alone, but they interact with their environment and it is this interaction that drives the development of our characters. In the previous chapter, I presented the concepts of developmental plasticity and developmental robustness. It seems that intuitively we tend to pay more attention to the latter, i.e. to the capacity of individuals of the same species to exhibit the general characteristics of their species irrespectively of the environment they live in. To many people, this seems to be all that heredity is about: from parents to offspring the same characters seem to be transmitted, no matter their environment.

However, as I have explained, this is just one side of the coin. Organisms are also characterized by developmental plasticity, i.e. individuals of the same species with the same genotype may produce different phenotypes during development as a response to local environmental conditions (see Bateson & Gluckman, 2011, pp. 4–5, 8). In other words, there is no point in distinguishing between any genetic and environmental contribution to the development of a phenotype, because both of them are important and any phenotype is the outcome of the interaction of the two. It is perhaps time to abandon altogether the notion of "gene action," and always think in terms of "gene interaction." This is why single genes can only account for character differences and not characters themselves. Genes do not invariably determine characters and disease; but characters are affected by changes in genes. Characters are also affected by changes in the environment or by the different environments in which an individual lives. Actually, in some cases the cellular and the organismal environment impact the expression of genes, as I explain in the next chapter.

Genomes Are More than the Sum of Genes

The predominant model until recently has been that DNA encodes proteins, with RNAs having only an intermediate (e.g. mRNA) or an auxiliary role (e.g. tRNA, rRNA) in the process of protein synthesis. As a result, for a long time the DNA that does not encode proteins – about 98 percent of human DNA – was described as "junk" DNA. The term is attributed to Susumu Ohno (1928–2000) who argued that more than 90 percent of human DNA has accumulated in the course of evolution without currently having any significant role. He concluded by noting that "Triumphs as well as failures of nature's past experiments appear to be contained in our genome" (Ohno, 1972). Nevertheless, "junk" does not necessarily mean "rubbish." As Sydney Brenner, who seems to have used the term in the 1960s, nicely put it: "[T]here are two kinds of rubbish in the world ... There is the rubbish we keep, which is junk, and the rubbish we throw away, which is garbage. The excess DNA in our genomes is junk, and it is there because it is harmless, as well as being useless, and because the molecular processes generating extra DNA outpace those getting rid of it" (Brenner, 1998, p. R669). In other words, "junk" may just mean "unused" and not "useless." But even in this case, "junk" DNA was usually considered as less important than protein-coding DNA.

When the sequence of the human genome was published (Lander et al., 2001; Venter et al., 2001), most people thought of the genome as consisting of words (genes) dispersed within meaningless text (the rest of the genome). To illustrate this, first consider the following text:

Nxyasjldoeplrptueijcmslswoprlcmaqwsxnmyflprtiifbvncqwieorflpxasnv
mortipqaqkfldoaybymvlcldosjahdbdncmvmvmvnsalsksfotospqloeruiotnv
mjcjsoapaisdahdjakflflforjgjksnxveurunlppgamqiroyxcritpgswermlgot pgf
sdyaxerioplsnformdorutheldmakingovndkylxfdrhitopwieorpscnyjrlsjrotjsqo
elpaymfhsleuropsnxmkdopvnmsfkmvlspytuqwodjglptsensendahdjenrvbc
nqqoepprkdalynxmshairoesnxmvslofeirtvbcnnvmclsprshnvmckdopwisns

najskdlviprpemvnclsodorpdshjlueirbdlmslprotndndjfkfvlvotjfjhsjslkflqr
utipsmxbvncdjflrorqagenesmodlpovbcncmdjcbsmxbvncmldoprfhglaskpeow
xnymvjdlaprtuoswweqiorptsynamsireplmbkdshajyleoqashjfllcmcabyncms
dalqwuriofmcayxnmvlpdsueirplqweuiropsmyxnahsjldeuiwoqsnadnaslkgf
kdospeortusjasmahjdklsweiolfpreueioqpwxcnmdhsjalaperiwxmvnclpsdhjfr
utioepwlqmsaqlsyxcnfjrutgkotpdlspqweuitpcpxansdjklsperuitombbxnask
ewuqieplorrfjkdxncmyasweuiolptifncmldspaqweruivmcahgfjkritlpldmcbs
nncmcldpeuqabdfhrtffghdsllapweudjmvndklpwezhf.

At first sight, this text seems to be totally meaningless. This is what DNA
sequencing produces: a sequence of letters and nothing more. Actually, any
DNA sequence is more repetitive and boring than the sequence shown,
because it consists of only four letters (A, T, G, C). Now imagine that this
is a DNA sequence (one of the two DNA strands), which contains the four
exons of the fictitious human gene *MSG* (making sense of genes). If we
search carefully, and we know what we are looking for, we will be able to
spot the exons of this gene in this sequence, as indicated here:

Nxyasjldoeplrptueijcmslswoprlcmaqwsxnmyflprtiifbvncqwieorflpxasnv
mortipqaqkfldoaybymvlcldosjahdbdncmvmvmvnsalsksfotospqloeruiotnv
mjcjsoapaisdahdjakflflforjgjksnxveurunlppgamqiroyxcritpgswermlgotpgfs
dyaxerioplsnformdorutheldmakingovndkylxfdrhitopwieorpscnyjrlsjrotjsq
oelpaymfhsleuropsnxmkdopvnmsfkmvlspytuqwodjglptsensendahdjenrvbc
nqqoepprkdalynxmshairoesnxmvslofeirtvbcnnvmclsprshnvmckdopwisns
najskdlviprpemvnclsodorpdshjlueirbdlmslprotndndjfkfvlvotjfjhsjslkflqrut
ipsmxbvncdjflrorqagenesmodlpovbcncmdjcbsmxbvncmldoprfhglaskpeow
xnymvjdlaprtuoswweqiorptsynamsireplmbkdshajyleoqashjfllcmcabyncms
dalqwuriofmcayxnmvlpdsueirplqweuiropsmyxnahsjldeuiwoqsnadnaslkgf
kdospeortusjasmahjdklsweiolfpreueioqpwxcnmdhsjalaperiwxmvnclpsdhjf
rutioepwlqmsaqlsyxcnfjrutgkotpdlspqweuitpcpxansdjklsperuitombbxnas
kewuqieplorrfjkdxncmyasweuiolptifncmldspaqweruivmcahgfjkritlpldmc
bsnncmcldpeuqabdfhrtffghdsllapweudjmvndklpwezhf.

This is a way to represent how the exons of the fictitious *MSG* gene
are dispersed among other sequences, just like the words "making,"
"sense," "of," and "genes" are dispersed among the other letters.

The phrase "making sense of genes" consists of 18 characters,
whereas the whole text above consists of 900 characters. Therefore,

the sequence of the *MSG* gene spans 2 percent of the whole sequence. Thus, the sequence shown previously is also an illustration of what we mean when we state that protein-coding genes represent 2 percent of the human genome. So what about the rest? Is 98 percent of our DNA meaningless, as in the text above? Is it really "junk," perhaps the relic of our evolutionary history during which DNA sequences were simply accumulated? The answer is no, and in this chapter I explain why. The relevant knowledge has been emerging during the recent years, and we have come to know that much of what we used to call "junk" DNA seems to have important functions, particularly in the regulation of the expression of genes. Therefore, when we talk about a genome we should not only have genes in mind because this would leave out most of its DNA. Rather, the genome should be conceived as the whole DNA included in chromosomes (Barnes & Dupré, 2008, pp. 76–77). However, it should be noted that although the details have emerged recently, several researchers had long been aware that "junk" DNA was not entirely useless and that some DNA that does not code for proteins has important roles (see Palazzo & Gregory, 2014).

Already in 1961, Jacob and Monod speculated that an RNA molecule could be the repressor of gene expression: "The experiments are negative, as far as the chemical nature of the repressor itself is concerned, since they only eliminate protein as a candidate. They do, however, invite the speculation that the repressor may be the primary product of the *i+* gene, and the further speculation that such a primary product may be a polyribonucleotide" (Jacob & Monod, 1961, p. 333). However, it was later shown that that repressor is a polypeptide (Gilbert & Muller-Hill, 1966). By that time, it had already been clear that nontranslated or noncoding RNA molecules, such as rRNA and tRNA, have an important role in gene expression (Palade, 1955; Hoagland et al., 1958). But as the ENCODE project showed, there are other functional sequences outside protein-coding genes, which encode certain noncoding RNA molecules. This led to the expanded definition of genes presented in Chapter 4, which includes the genes for noncoding RNA as well. Except for tRNA and rRNA, these genes encode other types of RNA molecules, such as small nucleolar RNAs (snoRNAs) that are involved in RNA editing and microRNAs (miR-NAs) that have important regulatory functions. Although the details are still under study, the emerging evidence suggests that there are a lot more genes encoding regulatory RNAs than proteins in the human genome (Morris & Mattick, 2014).

The researchers participating in the ENCODE project actually went as far as to claim that 80 percent of the genome is transcribed and produces RNA molecules. As stated in the abstract of the respective article in *Nature*: "The Encyclopedia of DNA Elements (ENCODE) project has systematically mapped regions of transcription, transcription factor association, chromatin structure and histone modification. These data enabled us to assign biochemical functions for 80 percent of the genome, in particular outside of the well-studied protein-coding regions." In particular, this report states that researchers integrated results from experiments with 147 different cell types and other resources, such as GWAS, and concluded: (1) the vast majority (80.4 percent) of the human genome participates in at least one biochemical RNA- and/or chromatin-associated event in at least one cell type; (2) an initial classification of the genome to different chromatin states indicated 399,124 regions with enhancer-like features and 70,292 regions with promoter-like features; (3) it was possible to correlate quantitatively RNA sequence production and processing with both chromatin marks and transcription factor binding to promoters; and (4) the majority of SNPs associated with disease in GWAS were residing in or near ENCODE-defined regions that are outside of protein-coding genes with the disease phenotypes being associated with a specific cell type or transcription factor in many cases (ENCODE Project Consortium, 2012).[1] These results suggest that most of the human DNA has some function, raising the amount forty times: from 2 percent to 80 percent. However, this suggestion has raised some harsh criticism, as it was perceived to imply that biological function could be maintained without natural selection (see Eddy, 2013, for an overview).

One criticism pointed out that, if one agreed with the ENCODE consortium, 80 percent of our genome is functional and less than 10 percent of it is evolutionarily conserved through negative (purifying) selection,[2] the resulting conclusion being that 70 percent of the genome is invulnerable to deleterious mutations, either because no mutation

---

[1] See also www.genome.gov/27549810/2012-release-encode-data-describes-function-of-human-genome/ and www.nature.com/encode/#/threads

[2] This is selection against deleterious mutations, which simply means that they are removed from the population because the organisms that have them die out.

can ever occur in these "functional" regions or because no mutation in these regions can ever be deleterious. The criticism focused on how the ENCODE researchers defined function. According to the critics, the argument of the ENCODE researchers was that DNA segments with biological functions (e.g. the regulation of transcription) tend to display certain properties (e.g. transcription factors bind to them). Therefore, if a DNA segment is found to have the same property, it can also be considered as functional. Critics argued that this argument is false because a DNA segment may display a property without necessarily being functional, as e.g. a transcription factor may bind to a DNA sequence without necessarily resulting in transcription (Graur et al., 2013). Function, according to the critics, makes sense only in the light of evolution, and the functionality of a certain DNA sequence may change along its course. Thus, in a subsequent article they proposed an evolutionary classification of genomic sequences for this purpose. They first divided the genome into functional DNA, which has a function that is the outcome of natural selection, and rubbish DNA, which does not have a function. They further divided each of these two categories into two more. Functional DNA was divided into literal DNA, in which the exact order of nucleotides is under selection, and indifferent DNA, in which only the presence or absence of the sequence is under selection. They also divided rubbish DNA into junk DNA, which does not affect the survival and reproduction of the organism and thus evolves without being selected, and garbage DNA, which affects the survival and reproduction of its carriers but for the moment exists in the genome. The important point that these authors made was that which of these categories a certain DNA sequence belongs to may change during evolution, and thus functionality should only be determined on its current status (Graur et al., 2015).

Another criticism made two important points, among others. The first is the importance of distinguishing between a function that is the outcome of natural selection in the recent past and that is under purifying selection now, and a function that is considered for its current perceived role without considering past or current selection. This is similar to the argument that points out that function is a historical concept that makes sense only if one considers past selection for its emergence. The problem in the case of the ENCODE project according to this critic is that the distinction between these two conceptions of function was

GENOMES ARE MORE THAN THE SUM OF GENES 213

not explicitly considered. A second point was that an essentialist way of thinking had been adopted by the ENCODE researchers that, according to this critic, encouraged: (1) the attribution of functions known for only a few DNA sequences to a whole class of such DNA sequences; and/ or (2) the attribution of a function known for a certain part of a DNA sequence to its whole length. The problem here is that such attributions may extrapolate, without any justification, the functional significance of a certain DNA sequence, and attribute it to many more although they might not have it. Finally, it was argued that several criteria under which a philosophically informed theoretical reconsideration would turn most of "junk" DNA into functional DNA are possible, but these are not found in the arguments of the researchers of the ENCODE project (Doolittle, 2013).

A response to these criticisms has been that the differential expression of RNAs, including extensive alternative splicing, is a far more accurate guide to revealing the functional sequences of the human genome than assessments of the conservation, or lack of conservation, of DNA sequences through selection. In addition, the evidence for the biological function of noncoding RNAs in different developmental and disease contexts are – according to the defenders of this view – adequate in order to draw broader conclusions about the possible functions of other similar sequences. Thus, these authors suggested that the criticism of the conclusions of the ENCODE researchers was not due to the lack of the relevant evidence, but due to other reasons. The first was the long-standing protein-centric conception of gene structure and regulation in human development that should now be reconsidered. In other words, what is needed in this case is a shift in perspective in order to consider the ENCODE findings under a new light, and not under the one in which research has been conducted since the 1940s. The second reason was that the idea of a largely nonfunctional genome full of "junk" DNA has been used as evidence for the evolutionary accumulation of DNA sequences and against any notion of intelligent design. Such an argument is apparently threatened if one accepts that 80 percent of the human genome is functional (Mattick & Dinger, 2013).

So, what conclusion can be made? It is hard to tell at this point. The traditional way to assess functionality has been to compare the DNA sequences of different species and document the changes. In the case of humans, functionality can be inferred by comparing human DNA sequences with the respective ones in model organisms. The main assumption of this approach is that functional elements should not be very different between different species. The reason for this is that changes may affect functionality in detrimental ways; therefore, it is unlikely for changes to have been conserved evolutionarily. In contrast, changes in nonfunctional parts of the genome may have no impact at all, and therefore may have been accumulated in a lineage in the course of evolution. If one applies these criteria, one finds conservation of protein coding sequences (up to 85 percent between humans and mice), but little conservation outside them. Overall, this suggests that only 3 percent of the whole genome is functional, but this might overlook species-specific functional sequences. Another way to assess functionality is by measuring biochemical activity. This was the approach taken by the ENCODE consortium, with which they concluded that 80 percent of the genome is functional. Apparently, it is difficult to reconcile the two approaches and decide what is the case. However, it should be noted that the ENCODE researchers used immortal cell lines for some of the analyses, i.e. cells that can divide indefinitely. These cells seem to be transcriptionally more active than others. Therefore, the high levels of transcriptional activity found by ENCODE researchers may be indicative of the features of these cells, and not of all cells in the human body (see chapter 10 of Parrington, 2015).

Beyond this debate, the important conclusion of the ENCODE project and the relevant research is that RNAs that do not code for proteins have very important roles in cells. There are many roles that we currently know, and it is very likely that several others will be found in the future. Parrington (2015, pp. 83–84) provides a useful classification of noncoding RNAS (for more details see Cech & Steitz, 2014; Morris & Mattick, 2014): (1) siRNAs (small interfering RNAs) that regulate gene expression by inactivating specific mRNAs; (2) miRNAs (micro RNAs) that prevent specific mRNAs from making proteins by inhibiting their interaction with the ribosome; (3) piRNAs (piwi-interacting RNAs) that protect the genomes of ova and sperm by suppressing the activity of transposons by tracking them down and disabling them; and (4) lcRNAs (long noncoding RNAs) that bring together different parts of the genome to interact. It is important to note that, whereas transcription factors regulate gene expression only at the transcription level, noncoding RNAs affect this process at multiple levels (see Holoch & Moazed, 2015).

Overall, the ENCODE data supported the conclusion that chromatin state and transcription are related, although there is still a lot to figure out (Stamatoyannopoulos, 2012). As described in Chapter 4, DNA is not "naked," but packed with proteins forming chromatin. There are several phenomena producing modifications in chromatin. These cause no actual change in the nucleotide sequence of genes, but they can strongly affect their expression, i.e. whether or not a protein or an RNA molecule is produced. All these modifications can be permanent and can even be transferred across generations. As this is inheritance of modifications that are not made in the DNA sequence but "upon" it, it is described as epigenetic inheritance (the prefix "epi-" literally means "upon" in Greek). The term is attributed to Conrad Waddington (1905–1975) who proposed it for the studies intending to discover the "causal mechanics at work" by which genes bring about phenotypic effects, and "to relate them as far as possible to what experimental embryology has already revealed of the mechanics of development." He also proposed the term "epigenotype" for the complex developmental processes that lie between genotype and phenotype, which also connect them to each other (Waddington, 1942; for the history of the term, see Jablonka & Lamb, 2002; Hall 2011).

The term epigenetics currently refers to all interactions between DNA and its local environment that eventually influence gene expression (Moore, 2015, pp. 20–22). More technically, the term epigenetics generally describes "chromatin-based events that regulate DNA-templated processes" (Dawson & Kouzarides, 2012). Therefore, "epi-" in the sense of "upon" has an almost literal meaning as all the relevant modifications are made upon DNA itself or on the neighboring histones. Some of these modifications can be permanent, as if a play director took the script of *The Phantom of the Opera* (see Chapter 9) and made notes on it with a pen. But some other modifications can be reverted, introduced again and then reverted again, and so on, as if a play director took the script, made notes with a pencil, erased them, made the same or new notes again, and so on. These modifications affect whether genes are activated or deactivated, and therefore expressed or not expressed, under different conditions. So, let us see what kind of modifications take place.

The first of these modifications is called DNA methylation. Simply put, methylation is the addition of a small chemical group called methyl ($-CH_3$) to cytosine (C). This is done by enzymes called DNA methyltransferases (DNMTs – as their name indicates they transfer methyl groups to DNA). Thus, a molecule called methyl-cytosine, or methyl-C, is produced. More specifically, the methyl group is added to a C that is followed by a G in the same DNA strand. The sequences consisting of several pairs of these two consecutive bases are described as CpG islands.[3] These are not randomly distributed in the genome but are usually found upstream (before) genes, within their promoters. When CpG islands are highly methylated, the respective genes are not expressed; in contrast, genes with a low level of methylation in CpG islands are expressed. The protein MECP2 (methyl-CpG binding protein 2) binds to highly methylated DNA and blocks the expression of the respective gene by attracting other proteins that also bind there or by not allowing RNA polymerase to produce mRNA. What is even more important is that the methylation pattern of a DNA molecule is preserved after DNA replication that precedes cell division, and thus the new cells inherit it (Bird et al., 1985; Lewis et al., 1992; Adkins & Georgel, 2010; Carey, 2012, pp. 56–60). This entails that whatever DNA sequence an individual has in a given gene, it may not make much of a difference if this gene is not expressed. The important idea here is that the script of *The Phantom of the Opera* does not only include information about who says what, but also about when one says anything and even whether one speaks at all. The director may use a pen to cross out some lines in the initial script, to simply put them within parentheses or to make some notes. In the case of DNA methylation, these changes can be permanent. Whatever the message in these lines, it makes no difference if these are crossed out and nobody reads them.

There are other types of changes that are less permanent, as if the play director used a pencil – not a pen – to cross out lines, put them within parentheses, or write notes. These are changes in the histone molecules, which have an impact on gene expression. The first change of this kind to be identified was histone acetylation, i.e. the addition of an acetyl group (CH_3CO-) to amino acid lysine in particular histones. This seemed to facilitate the binding of DNA transcription factors and

---

[3] The "p" in CpG refers to the phosphodiester bond between C and G. This is the kind of chemical bond that connects any two nucleotides in the same strand.

thus to result in increased expression of nearby genes (Brownell & Allis, 1996; Vettese-Dadey et al., 1996; Wolffe & Pruss, 1996). Several other changes like this, which either disrupt chromatin or affect the binding of nonhistone proteins to it, can affect the structure of chromatin and thus whether enzymes can access DNA (Kouzarides, 2007; see also Carey, 2012, pp. 65–74). One example is histone methylation, i.e. the addition of a methyl group ($-CH_3$) to a specific lysine of a particular histone (Ng et al., 2010). What is even more interesting is that histone modification and DNA methylation seem to be related phenomena, as the former seems to affect the occurrence of the latter. For example, in mammals, DNA methylation and methylation of lysine at position 9 of histone H3 are strongly associated, whereas methylation of lysine 4 of histone H3 seems to inhibit DNA methylation (Du et al., 2015). Therefore, histones are more than just a means of packaging DNA, as histone modifications can affect DNA methylation and in turn gene expression. Which epigenetic marks (DNA methylation, histone methylation, or histone acetylation) a cell contains (Figure 11.1), in other words the overall epigenetic state of a cell, is described as its epigenome. These modifications seem to form a special, complex code that is "read" by proteins, and that in turn affects the accessibility and function of eukaryotic genomes (Rothbart & Strahl, 2014).

An effect of epigenetic modifications is easy to observe in calico cats, i.e. cats that have a coat with a color that is a mixture of black and ginger spots. You can impress your friends by guessing that these cats are females just by looking at their coat color. What happens in this case is that the gene that affects coat color is found on X chromosomes. A female cat can be heterozygous and carry an allele that is implicated in the development of black color (B) and another allele that is implicated in the development of ginger color (G); therefore it can have the genotype X^BX^G. However, in female cats one of the X chromosomes becomes inactivated. Thus, both males and females actually have a single, active X chromosome in each cell. Now what happens to female calico cats is that in some cells the X^B chromosome is inactivated, whereas in other cells it is the X^G that is inactivated. As a result, the coat of the cat becomes a mosaic of colors because it is a mosaic of cells with different inactivated X chromosomes, expressing either the B or the G allele (Carey, 2012, pp. 170–171; Moore, 2015, pp. 48–50).

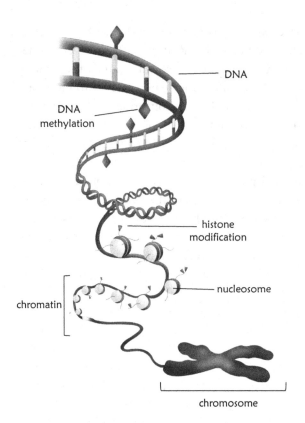

FIGURE 11.1 The two main kinds of epigenetic marks: DNA methylation and histone modification. In DNA methylation methyl groups are added to certain cytosines. In this way, gene expression is repressed. In histone modification, a combination of different molecules can attach to the "tails" of histones. These alter the activity of the DNA wrapped around them (© Media for Medical).

A relevant but different phenomenon is that of genomic, or epigenetic, imprinting. In some cases, the chromosomes of one of the twenty-three pairs that each of us has in his/her cells cannot be distinguished from each other. There is nothing that indicates which of the two homologous chromosomes was inherited by one's mother and which one was inherited by one's father. However, for some chromosomes such as chromosome 11 this is possible. Mammals have evolved to have epigenetic marks that distinguish between the parental and the maternal chromosome. But this is not just a distinction issue; maternal and paternal chromosomes are not always equivalent. In the early 1980s, nuclear transplantation experiments had shown that both parental genomes are necessary for a complete development. An interesting finding emerged in experiments with mice that had inherited one chromosome of each pair from each of their parents, except for chromosome 11. In this case, some mice had inherited either two paternal chromosomes 11, or two maternal chromosomes 11 (this condition is described as uniparental disomy). The observed outcome was a consistent difference in the size of these mice. Mice having both a paternal and a maternal chromosome 11 were normal in size; but mice with chromosomes 11 of maternal origin were smaller than the others, whereas mice with chromosomes 11 of paternal origin were larger than the others (this was the case for both male and female mice) (Cattanach & Kirk, 1985; Cattanach, 1986).

These results showed that maternal and paternal chromosomes are not functionally equivalent. But why is this happening? It seems that certain genes that are switched on in one chromosome are switched off in the other chromosome. Therefore, when an organism has two chromosomes of the same kind, there is hyperactivity or hypoactivity of these genes. In the case of the mice described above, genes related to placental growth were very active in the paternally derived chromosomes and not very active in the maternally derived chromosomes (Carey, 2012, p. 134). This "switching on" and "switching off" of particular genes of the same chromosome is due to epigenetic marks that exist on the chromosome of the one parent but not on that of the other. This phenomenon is described as genomic imprinting. During this process, the alleles of particular genes acquire an epigenetic mark (e.g. DNA methylation) at specific sequences called imprinting control regions. As a result, only one of the alleles of these genes is expressed in the offspring, and these are distinguishable in terms of parental or maternal origin. This imprinted gene expression is maintained by various mechanisms such as noncoding RNAs, and histone modifications (Bartolomei, 2009). Recent research has shown that genomic imprinting varies between genes, individuals, and tissues (Baran et al., 2015). An interesting instantiation of this has been observed for ages. It has been long known that mating a female horse and a male donkey produces a mule, whereas mating a male horse and a female donkey produces a hinny. Hinnies are a bit smaller than mules, and they also differ in their physiology and temperament. It seems that these differences are due to differences in paternally expressed genes as a result of imprinting (Wang et al., 2013).

The most well-known examples of genomic imprinting in humans are the Prader-Willi and Angelman syndromes (Prader et al., 1956; Angelman, 1965). In both cases, the syndromes are associated with the deletion of the same part of chromosome 15, and what makes the difference is whether the X chromosome with the deletion was inherited from the mother or from the father (Knoll et al., 1989). In particular, the Prader-Willi syndrome is associated with the deletion on the paternal chromosome 15, whereas the Angelman syndrome is associated with the deletion on the maternal chromosome 15. In people with Prader-Willi, the paternal chromosome has the deletion, and the respective section on the maternal chromosome is imprinted and thus unexpressed. However, it is not the deletion that causes the problem, as people with Prader-Willi have been found to have two complete chromosomes 15. It seems that these are two maternal chromosomes that are imprinted and thus nonexpressed (Nicholls et al., 1989). In the case of the Angelman syndrome, the deletion exists on the maternal chromosome 15 and it is the paternal chromosome 15 that is imprinted and unexpressed. From these observations, it becomes clear why these syndromes are both genetic and epigenetic in nature. The effects of a chromosome deletion are not compensated by the extant sequences in the complementary chromosome because the latter are imprinted and thus switched off (Carey, 2015, pp. 136–141; Moore, 2015, pp. 46–48). Furthermore, where this happens makes a difference to which syndrome these people will develop. This is because the genes involved in each case are not exactly the same. The Prader-Willi syndrome is due to loss of paternal expression of up to eleven genes as a result of the deletion in the paternal chromosome. The Angelman syndrome arises from loss of maternal expression of the *UBE3A* (ubiquitin protein ligase E3A) gene, mainly due to the deletion in the maternal chromosome (Lee & Bartolomei, 2013; Peters, 2014).

An even more interesting finding is that epigenetic differences can account for phenotypic differences between monozygotic twins, who are otherwise supposed to have (almost) the same DNA sequences. A study with eighty monozygotic twins showed that although they were indistinguishable during their early years of life in terms of DNA methylation and histone acetylation, in older twins there were remarkable differences in the genomic distribution of these marks that affected gene expression. These differences were more significant between older monozygotic twins who had different lifestyles, and had spent less of their lives together, thus highlighting the significant impact of environmental factors on the development of different phenotypes on the basis of the same genotype (Fraga et al., 2005). Interestingly, although the differences are small, there is variation in DNA methylation even between newborn monozygotic twins, as well as between different tissues. Such results suggest that the intrauterine period is a crucial period for the establishment of epigenetic marks in humans (Ollikainen et al., 2010). This means that no matter what DNA or gene sequences one has, knowing it is not enough for predicting the phenotype. Which epigenetic marks one has and why also makes a crucial difference.

Epigenetic marks are so important that they can actually determine the potency of a cell, which can be:

1. Totipotent, i.e. divide and eventually produce all the cell types of the body and the placenta. The blastomeres of the early embryo are totipotent cells.
2. Pluripotent, i.e. divide and eventually produce all the cell types of the body but not the placenta. The cells of the inner cell mass of the blastocyst are pluripotent cells.
3. Multipotent, i.e. divide and eventually produce various cell types of the body. Most adult stem cells, such as skin stem cells, hematopoietic stem cells, and neural stem cells are multipotent.
4. Unipotent, i.e. divide and eventually produce only one cell type. Examples of such cells are differentiated cells, certain adult stem cells (e.g. testis stem cells), and certain progenitor cells (e.g. erythroblasts) (based on Hochedlinger & Plath, 2009).

Waddington (1957, p. 29) came up with a very effective illustration of these differences: the epigenetic landscape. According to this, the embryo can be thought of as an undulating plateau. On the top of it there is a ball that corresponds to a single cell. Below, there are several possible trajectories leading to different positions; these correspond to the different developmental paths that a cell can take and that through which it will develop to one or another type of cell (Figure 11.2). Pluripotent cells are to be found on the top of the plateau and the fully differentiated ones at the end of each trajectory, with the various intermediate types in between (Hochedlinger & Plath, 2009, p. 510). What is now clear is not only that the cells of various kinds of potency are in

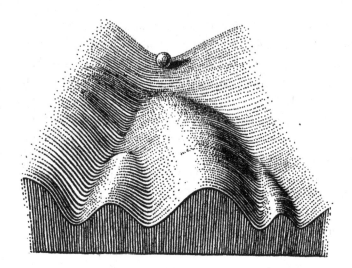

FIGURE 11.2 Waddington's epigenetic landscape. This visual metaphor can be used to represent the process of cellular differentiation during development. A cell can be represented as a ball that can take specific permitted trajectories, leading to different outcomes. Across the landscape, there are different degrees of differentiation and potency. Totipotency would be on the top of the figure, where the ball is found; pluripotency would be a bit lower, before the different paths begin; multipotency would be in the beginning of the two paths on the left; and unipotency would be at the end of each path. From Waddington, C. H., The *Strategy of Genes*, © 1957, Allen & Unwin. Reproduced with permission from Taylor & Francis Books, UK.

a different epigenetic state, but also that differentiated cells can revert to less differentiated states if their epigenetic state changes. Adding particular transcription factors can turn somatic cells to a pluripotent state without the use of oocytes, generating what has been called as induced pluripotent stem (iPS) cells (Takahashi & Yamanaka, 2006; Yamanaka, 2009).

It is important to note at this point that whereas epigenetic modifications are generally stable in the somatic cells of mammals, all or most (this is not entirely clear yet) of the epigenetic marks (DNA methylation, histone acetylation, histone methylation, etc.) are erased and are created anew on a genome-wide scale at two distinct stages: in the primordial germ cells (PGCs) when they have reached the embryonic gonads, and in the early embryo between the stage immediately after fertilization and the blastocyst stage. This process is described as epigenetic reprogramming, and it contributes to resetting the epigenetic marks. This means that epigenetic marks are not transmitted unchanged across generations; they are recreated in each new one. The process of epigenetic reprogramming has important roles in imprinting, in the acquisition of totipotency and pluripotency, and in the expression of transposons or other potentially "dangerous" DNA sequences (Feng et al., 2010). For example, erasure of the epigenetic marks after fertilization is necessary in order for the cells of the early embryo to acquire pluripotency.

Let us see in some detail how this process takes place. The first phase of epigenetic reprogramming is in the early embryo. The paternal genome in the zygote is rapidly demethylated, probably through an active process, whereas the maternal genome becomes demethylated more slowly, perhaps in a passive process.[4] Then, remethylation takes place and occurs during the early development of the embryo. It seems that imprinted genes do not undergo this kind of epigenetic reprogramming. The second phase of reprogramming takes place in PGCs. During this phase, even imprinted genes are demethylated and get their new marks according to the sex of the offspring. It is empirically known that some epigenetic marks escape reprogramming at both phases. We also know that there are specific mechanisms that retain these marks in the early embryo, as is evident from imprinted genes that are consistently methylated here. Therefore, it is possible for transgenerational, epigenetic inheritance to occur. Apparently, this requires different kinds of proteins involved in epigenetic signaling. Some deposit epigenetic marks ("writers"), some others remove epigenetic marks ("erasers"), whereas a third category are those proteins that "interpret" these marks and initiate the respective regulatory process ("readers") (see Seisenberger et al., 2013; Heard & Martienssen, 2014; Hughes, 2014; von Meyenn & Reik, 2015) (Figure 11.3).

Epigenetic modifications have an important impact on cancer (see also Chapter 7). One of the first epigenetic modifications found in human cancers was the low level of DNA methylation in tumors compared to the respective "normal" tissues.[5] DNA hypomethylation can produce chromosomal instability by promoting mitotic recombination, which

---

[4] It is not yet clear how active demethylation takes place; it could be a demethylating enzyme. Passive demethylation takes place when DNA methylation patterns change during DNA replication because the bases are not methylated anew in the new strands (Moore, 2015, pp. 72–73).

[5] For the history of cancer epigenetics, see Feinberg and Tycko (2004).

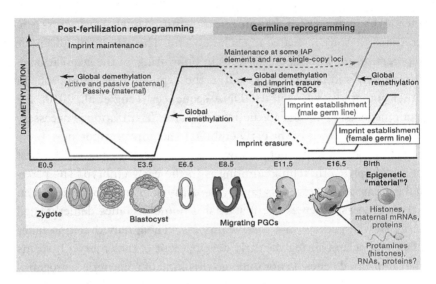

FIGURE 11.3 In mice, there are at least two rounds of genome-wide DNA methylation reprogramming. The first occurs just after fertilization, in the zygote and early cleavage stages, and erases epigenomic marks that existed in sperm and oocyte. During this phase of reprogramming, genomic imprints are maintained. The other major reprogramming process occurs in the germline, where the paternal and maternal somatic programs are erased, together with imprints, and the inactive X is reactivated. Subsequent to this, parent-specific imprints are laid down in the germline. Reprinted from Heard, E., & Martienssen, R. A., "Transgenerational epigenetic inheritance: myths and mechanisms," *Cell*, 157(1), 95–109, © 2014, with permission from Elsevier.

leads to deletions and translocations, and chromosomal rearrangements. It can also reactivate transposable elements that can be transcribed or translocated to other parts of the genome, thus further disrupting it. Finally, DNA hypomethylation can disrupt genomic imprinting, and this can lead to cancer development. Another epigenetic modification found in cancers is the hypermethylation of CpG islands, which results in the inactivation of tumor-suppressor genes. Hypermethylation of CpG-islands in the promoter of genes involved in the cell cycle, DNA repair, the metabolism of carcinogens, cell-to-cell interaction, apoptosis, and angiogenesis can also impact the development of cancer. Finally, particular histone modifications have been found in a variety of cancers. There are several kinds of epigenetic modifications in several types of tumors (Esteller, 2008, 2011; see Figure 11.4).

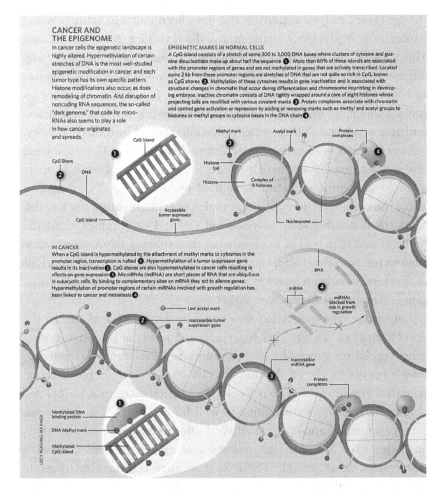

FIGURE 11.4 Epigenetic modifications in cancer. Reprinted from Esteller, 2011, with the kind permission of the author.

More recent research has shown that there are a significant number of cancer mutations that affect epigenetic pathways, which regulate modifications of DNA, histones, and chromatin remodeling. First, there exist mutations that affect DNA methylation by affecting the DNMTs, the enzymes that add methyl groups to cytosines. For example, mutations have been found in DNMT3A in acute myeloid leukemia, myeloproliferative diseases, and myelodysplastic syndromes. Mutations affect either the catalytic activity of the enzyme, or its localization to chromatin. Second, there exist mutations that impact histone acetylation by affecting histone acetyltransferases. Mutations in these enzymes may change their catalytic activities, and have been found in acute myeloid leukemia, acute lymphoid leukemia, and diffuse large B-cell lymphoma. Third, there are mutations that impact histone methylation by significantly altering the catalytic activity of histone methyltransferases and demethylases. These mutations have been found in a variety of cancers such as acute myeloid leukemia, diffuse large B-cell lymphoma, and multiple myeloma. Fourth, other mutations impact histone phosphorylation by affecting signaling molecules such as kinases. These are some of the most frequent oncogenic events found in cancers, such as acute myeloid leukemia and acute lymphoid leukemia. Finally, mutations affect the chromatin-remodeling complex that binds chromatin and disrupts the bonds between histones and DNA. This complex alters the position and structure of nucleosomes thus making DNA more accessible to transcription factors and other chromatin regulators. Mutations of this kind have been found in a variety of cancers, most notably breast cancer (Dawson & Kouzarides, 2012).

Epigenetic reprogramming also seems to have a crucial role in the development of cancer. This is actually the result of the activation of gene regulatory processes in inappropriate developmental contexts, which results in the production of cells with unlimited proliferation potential similar to that of iPS cells. Several transcription factors and chromatin regulators correspond to well-known oncogenes and tumor suppressor genes, and have established roles in oncogenesis. Therefore, analogies exist between the processes of epigenetic reprogramming and cancer development. These suggest that certain alterations in chromatin regulators in cancer may be early events that render a cell susceptible to epigenetic reprogramming, and thus open to potential differentiation. Subsequent changes, taking place in a variety of orders and combinations, may eventually turn a cell into a cancer one (Suvà et al., 2013). Cancers can start either from stem cells or from differentiated cells that become dedifferentiated. In both cases hyperproliferation and senescence bypass take place. Senescence is an important barrier to oncogenic transformation and the generation of iPS cells, and its bypass is promoted by MYC (the product of an oncogene) and the loss of tumor suppressor factors such as p53. MYC also promotes proliferation and reprograms metabolism. Reprogramming of mouse embryonic fibroblasts to iPS cells requires a mesenchymal-to-epithelial transition, i.e. a process during which motile mesenchymal cells are converted to polarized epithelial cells (i.e. cells with a certain up-down polarity). This is a process that cancer cells also undergo, which enables colonization of other tissues. Before that, however, primary tumors usually first undergo the opposite process, an epithelial-to-mesenchymal transition, during which they adopt an invasive phenotype. The significant parallels between the molecular mechanisms underlying reprogramming events leading to the generation of iPSCs and those that lead to the development of cancers suggest that cancers originating from stem cells may be due to mutations in genes that are implicated in the plasticity of DNA methylation (Goding et al., 2014).

The main conclusion from all these is that both genetic and epigenetic phenomena are related to genes but in very different ways. Genetic phenomena are related to the sequence of genes (coding or regulatory sequences), and phenotypic changes occur only as a result of a change in the base sequence of DNA. Such changes cannot happen because of environmental influences, except in the case of mutations. In contrast, epigenetic phenomena are related to chemical modifications of gene-related sequences that are not (at least considered as) genetic, as well as to similar modifications of the surrounding histone proteins. These modifications are very sensitive to the external environment. Although epigenetic phenomena do not entail the transfer of any information, they affect the expression of the information encoded in DNA. This makes necessary a reconceptualization of the central dogma (Figure 3.11) that also takes into account the effect of transcription factors and epigenetic phenomena. Actually, this should not be called a dogma at all, exactly because it can undergo changes, as with all scientific knowledge. The term "dogma" is very inappropriate exactly because dogmas are not

FIGURE 11.5. The postgenomic "central dogma." The gray arrows do not represent any flow of information, like the black ones do, but rather influences that affect the flow of information represented by the black arrows. The relative sizes of arrows are also indicative that the flow of information from DNA to RNA is larger than all the others, as active RNA transcripts are more numerous than proteins. This figure encompasses the complexity of DNA expression better than the typical representation found in textbooks, as it also shows the important effect of transcription factors and noncoding RNAs.

supposed to change (Crick later acknowledged the problems with the choice of this term; see Crick, 1988, p. 109). Nevertheless, for historical reasons we can retain the term "central dogma," and based on the research presented in this chapter propose its postgenomic version as in Figure 11.5. In my view, this highlights the importance of recent findings of genomics research. The importance of the phenomena described as epigenetic are still debated, but there is certainly something there. What is certainly necessary to show is the influence of proteins in the expression of genes, most notably the role of transcription factors and of noncoding RNA. Genes are not autonomous entities but are subject to control and regulation of expression that depends on environmental signals, as well as numerous proteins and RNA molecules.

So far, we have seen that genes are not segments of DNA that encode proteins and in turn determine characters. If genes do anything at all, it is that they are resources that encode information for the synthesis of RNA or protein molecules that are implicated in the development of characters. We have also seen that it is not the action of single genes but the interaction of many genes with one another and with their environment that drives the development of characters. Furthermore, several regulatory DNA sequences also seem to play a major role in the expression of genes and the development of characters, among complex molecular interactions within and among cells. The question now becomes: is this knowledge useful for medical purposes? Do genes really predispose us to certain diseases? Is it helpful to have a genetic test in order to know one's alleles related to some disease in order to take some precautionary or therapeutic action? Most importantly, how sure can we be about what we can know from DNA analyses? Are there any limitations that we may overlook? What should we anticipate that we can know in the end?

Francis Collins, director of the Human Genome Project, has suggested that knowledge about our genes and DNA can be really useful: "Identifying human genetic variations will eventually allow clinicians to subclassify diseases and adapt therapies to the individual patient. There may be large differences in the effectiveness of medicines from one person to the next. Toxic reactions can also occur and in many instances are likely to be a consequence of genetically encoded host factors. That basic observation has spawned the burgeoning new field of pharmacogenomics, which attempts to use information about genetic variation to predict responses to drug therapies" (Collins, 1999, p. 33). You will think, "OK, that was 1999, before the completion of the Human Genome Project. Perhaps it was too early back then and one could be optimistic." But here is Collins again after the first conclusions of the GWAS: "We are on the leading edge of a true revolution in medicine, one that promises to transform the traditional 'one size fits all' approach into a much more powerful strategy that considers each individual as unique and as having special characteristics that should guide an approach to staying healthy. Although the scientific details to back up these broad claims are still evolving, the outline of a dramatic paradigm shift is coming into focus" (Collins, 2010, pp. xxiii–xxiv). In the latter case, Collins certainly acknowledges that we are not there yet, but we have entered the era of personalized, genomic medicine.

Collins is of course right that people respond differently to diseases and therapies, and that therefore knowing the physiological peculiarities of each individual is important. Personalized medicine is unquestionably important. What is questionable is whether knowledge about genes and genomes currently makes any difference. This brings us to the first terminological distinction that is necessary to make: genomic medicine and personalized medicine are not synonyms. Genomic medicine is about the impact that one's genes have on disease. Personalized medicine is about taking individual characteristics into account, and these are not restricted to one's genomic information; these also include one's lifestyle, the environment in which one lives, and a lot more. Therefore, treating the terms as synonymous is misleading because it is like implicitly accepting that our characters can be reduced to our genes (Annas & Elias, 2015, pp. 25–26). My aim in this chapter is to explain the potential and the limitations of genomic medicine and of DNA analyses more generally. Collins concluded his book-length treatise of the topic by suggesting: "Your DNA helix, your language of life, can also be your textbook of medicine. Learn to read it. Learn to celebrate it. It could save your life" (Collins, 2010, pp. 278–279). Let us see how realistic this currently is.

At first sight, this suggestion seems plausible. "Why not?" you think. "I will take a DNA test. If it reveals something in my DNA that makes me susceptible to a disease, I will take all the necessary actions. If not, then even better!" Prevention is better than cure, as Hippocrates is said to have put it. Why wait until a disease develops and then fight against it? Why not prevent it before it even develops? As explained in Chapters 7 and 8, DNA tests can be used as the basis for estimating the risk for developing a disease. On the basis of such a knowledge your physician could then guide you to the appropriate actions that will help you prevent the disease, or at least fight it more efficiently.

Reading our genomes can provide us with valuable information, so it is better to know than live in ignorance – isn't it? However, notwithstanding the problems with the interpretations of genetic tests discussed in Chapters 7 and 8, many of which can be resolved if an expert guides one to understand their results, there are other issues to consider.

According to Eric Lander (1957–), one of the major figures in the HGP, a recent estimate is that there exist "3600 genes for rare mendelian disorders, 4000 genetic loci related to common diseases, and several hundred genes that drive cancer" (Lander, 2015, p. 1185).[1] How many of these do you really want to be tested for? How much knowledge of this kind is enough? Worse than that, with these numbers it is very likely that each one of us will carry several variants associated with a disease. In that case, how can we decide whether one is "healthy" or a "carrier," and the limit between the two conditions? To make such decisions, one should know which of these genes or loci are actually doing harm and which are harmless. This is presently very unclear for many of these. Lander suggests that regulation is necessary, pertaining to two very important aspects: analytical validity and clinical validity (see Chapter 8). According to Lander, for gene-based tests these aspects depend on the answers to the following two questions: "Does the test accurately read out a targeted set of DNA bases in the human genome? Does the targeted set of DNA bases provide meaningful clinical information?" (Lander, 2015, p. 1186). This could be possible to achieve one day but, as I explained in Chapters 7 and 8, we are not there yet!

Interestingly, Lander wrote these lines as a response to an article coauthored by Francis Collins that presented initial ideas about the implementation of the Precision Medicine Initiative announced by President Obama in 2015 (Collins & Varmus, 2015). Assuming that the various studies are successfully replicated and that the companies providing genetic tests clearly show what they are measuring and how, it could be possible to address the first question. The second question, that directly involves you and me and every interested citizen, is more difficult to address and this is the one on which I will focus in this chapter. Strange as it might sound, to address this question one does not only need a good understanding of genetics, but also a good understanding of probabilities. The necessary genetics knowledge has been described in the

---

[1] For updated numbers, see www.omim.org/statistics/entry.

previous chapters of this book. Finding an association between a DNA variant and a disease does not necessarily entail that someone having the variant will develop the disease. There are various reasons for this, including that different genes contribute to one's overall risk to develop a disease, with each one making only a minor contribution; that interactions among genes produce different combinations with different phenotypic outcomes, and so diseases are heterogeneous; and finally that the interaction between genes and environment is always crucial especially in the case of complex disease. Therefore, knowing one's genes (or DNA variants) associated with a disease is just one piece, or a few pieces, of the puzzle. An additional problem is that the actual knowledge that a test can confer are usually presented in a misleading way.

How about a concrete example? The lifetime risk for a woman to develop breast cancer is 12.3 percent. Lifetime risk, as also explained in Chapter 8, refers to the probability that an individual will develop cancer over the course of a lifetime. What does the probability 12.3 percent for a woman to develop breast cancer mean? Initially, perhaps not much. We saw in Chapter 8 that numbers like this make little sense to people, and that it is better to use frequencies rather than probabilities when talking about risks (Gigerenzer, 2002, chapter 4). Given this, the estimated risk for cancer of 12.3 percent can be described as approximately one in eight. What does this practically mean? That one in eight women will develop, and might die from, breast cancer. A woman reading these lines may now think of other seven women whom she knows and conclude: "One of us will die from breast cancer!" Yes, but note: this is a lifetime risk! Therefore, the number does not indicate that one woman out of eight will die from breast cancer at any age. Unfortunately, death is the only certain future outcome once we are born, and so the probability for any individual to die at some point in life is 100 percent. Given this, what this number suggests is that one woman out of eight will develop breast cancer and might die because of it, whereas the other seven will die from something else. It is also important to note that this is just statistics, not some law with predictive power. This means that it is possible that no woman within a particular group of 8 women will develop breast cancer. Let us look into actual numbers in more detail to make this clearer. Table 12.1 presents the estimated risk for women to develop breast cancer in various ages (American Cancer Society, 2016, p. 17). These risks are presented in a slightly different way in Table 12.2.

TABLE 12.1 *Age-Specific Probabilities of Developing Invasive Female Breast Cancer*

Current age	Probability of developing cancer in the next 10 years	
20	0.1%	1 in 1,674
30	0.4%	1 in 225
40	1.4%	1 in 69
50	2.3%	1 in 44
60	3.5%	1 in 29
70	3.9%	1 in 28
Lifetime	**12.3%**	**1 in 8**

Source: Data from Table 5, American Cancer Society (2016, p.17).

TABLE 12.2 *Age-Specific Probabilities of Developing Invasive Female Breast Cancer*

Current age	Number of women with breast cancer	Total number of women	Probability of developing cancer in the next 10 years
20	1	1,674	0.1%
30	1	225	0.4%
40	1	69	1.4%
50	1	44	2.3%
60	1	29	3.5%
70	1	28	3.9%
Lifetime	**1**	**8**	**12.3%**

Source: Adapted from Table 5, American Cancer Society (2016, p.17); see also Hubbard and Wald (1997, p. 87); Gigerenzer (2002, pp. 76–78).

Now, how much is the impact of genes in the development of breast cancer? Consider the following statements in the report of the American Cancer Society (2016, p. 11): "Compared to women in the general population who have a 7% risk of developing breast cancer by 70 years of age, the average risk for *BRCA1* and *BRCA2* mutation carriers is estimated to be between 57%–65% and 45%–55%, respectively." What does this mean? It means that, whereas by 70 years of age 7 out of 100 women of the general population will have developed breast cancer, 57–65 out of 100 *BRCA1* mutation carriers and 45–55 out of 100 *BRCA2* mutation carriers will develop breast cancer by the same age. This means that, whereas less than one out of ten women from the general population will develop cancer by the age of seventy, for women carrying a relevant *BRCA1* or *BRCA2* mutation it will be approximately one out of two. This initially sounds a lot. However, earlier in the same paragraph it is stated: "Inherited mutations (genetic alterations) in *BRCA1* and *BRCA2*, the most well-studied breast cancer susceptibility genes, account for 5%–10% of all female breast cancers, an estimated 5%–20% of male breast cancer, and 15%–20% of all familial breast cancers." This means that out of the 100 women who will develop breast cancer, only 5–10 will be *BRCA1* and *BRCA2* mutation carriers.

In other words, 90–95 of the women who will develop breast cancer will not be *BRCA1* and *BRCA2* mutation carriers! What is the conclusion? Women carrying *BRCA1* and *BRCA2* mutations should be concerned, but these are only a minority of the women developing breast cancer. The extreme focus on *BRCA1* and *BRCA2* mutations, especially through public accounts such as the Angelina Jolie story discussed in Chapter 5, perhaps attract the attention from other factors that affect more women than these alleles do. As the report of the American Cancer Society notes on the same page: "These mutations are very rare (much less than 1%) in the general population, but occur slightly more often in certain ethnic or geographically isolated groups, such as those of Ashkenazi (Eastern European) Jewish descent (about 2%)." This means that approximately 1 out of 100 women will have a *BRCA* mutation and develop cancer. But this is not always emphasized. Figure 12.1a presents an account of the impact of *BRCA1* and *BRCA2* mutations on cancers that we find frequently (e.g. www.cancer.gov/news-events/cancer-currents-blog/2016/brca-testing-breast-cancer). This is relevant only for women who are carriers of a *BRCA* allele. Figures 12.1b and 12.1c present an account that is relevant to all women.

If you look at Figure 12.1c carefully, there are two important take-home messages: (1) 93 out of 100 women will not develop breast cancer by the age of 70; (2) 6 out of 7 women who will develop breast cancer will not carry any *BRCA* mutation. I think that these figures should be considered before having a genetic test about a *BRCA* allele, and a woman could perhaps be skeptical about the utility of such a test. If there is a

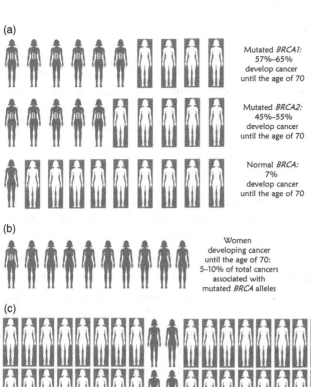

(a)

Mutated *BRCA1*:
57%–65%
develop cancer
until the age of 70

Mutated *BRCA2*:
45%–55%
develop cancer
until the age of 70

Normal *BRCA*:
7%
develop cancer
until the age of 70

(b)

Women
developing cancer
until the age of 70:
5–10% of total cancers
associated with
mutated *BRCA* alleles

(c)

Women with mutated
BRCA alleles
developing cancer
until the age of 70:
5–10% of 7%
of the general
population, i.e.
less than 1%
of the general
population

Legend

woman with cancer
and mutated *BRCA*
alleles

woman with cancer
but without mutated
BRCA alleles

woman without cancer

FIGURE 12.1 (a) The usual representation associating risk for developing breast cancer with *BRCA* alleles. (b) A representation showing that a minority of women who develop breast cancer actually have the mutated *BRCA* alleles. (c) A representation showing that at most 1 out of 100 women will develop breast cancer and carry the mutated *BRCA* alleles (Character outline © filo).

history of breast cancer in the family, then a woman should consider things differently than a woman without a family history. Of course, this way of considering probabilities makes no difference to someone who eventually develops the disease. However, the possible risks should be evaluated in a way that is as objective as possible (Hubbard & Wald, 1997, p. 28). This brings us to the important distinction between genetic screening and genetic diagnosis. Screening refers to looking for conditions or specific alleles in the general, healthy, and asymptomatic population. Given that at most 1 out of 100 women will carry *BRCA* alleles and develop breast cancer until the age of 70 years old, one should wonder whether screening for these alleles makes any sense. This is very different from diagnosis, i.e. looking for conditions or specific alleles to people already considered as likely to have them, such as people with a family history of breast cancer. In the case of *BRCA* alleles, attempting to confirm or disconfirm the presence of a condition or an allele in people with a family history of breast cancer would make sense (see Annas & Elias, 2015, p. 116). It is one thing to recommend that a woman with a family history of breast cancer should undergo diagnosis for *BRCA* alleles, as they are clearly associated with the disease (e.g. Easton et al., 2015) and another that all women should be screened for these alleles. Other methods of screening such as mammography also do not have the clear-cut results that one would expect (Pace et al., 2014).

But there is more that further complicates the picture. A recent headline poses the problem (in what seems to me a surprised mood): "Puzzle of seven sisters who've all had breast cancer ... and tested NEGATIVE for BRCA genes."[2] Is this really a puzzle? In a study of 198 women, 174 of which had breast cancer, several pathogenic variants and variants of unknown significance (VUS) outside the *BRCA1* and *BRCA2* genes were identified. In particular, the entire coding sequences, exon-intron boundaries, and all known pathogenic variants in other regions for forty-two genes that had cancer risk associations were sequenced. Fifty-seven women were found to carry thirty-five *BRCA1* and twenty-four *BRCA2* mutations. But pathogenic variants in 9 genes[3] were also identified in

---

[2] www.dailymail.co.uk/news/article-3308847/Puzzle-seven-sisters-ve-breast-cancer-tested-NEGATIVE-negative-BRCA-genes.html

[3] These genes were *ATM, BLM, CDH1, CDKN2A, MUTYH, MLH1, NBN, PRSS1,* and *SLX4.*

16 among the 141 women without any *BRCA1* and *BRCA2* mutations. These results indicate that mutations in *BRCA1* and *BRCA2* genes were found in approximately 27 percent of the participants, whereas other pathogenic variants were found in approximately 8 percent of the participants, and in approximately 11.5 percent of those participants without any *BRCA1* or *BRCA2* mutations. Furthermore, the sequencing of forty-two genes allowed the identification of variants of unknown significance in many participants (88 percent), with an average of 2.1 VUS per participant. Such results show that it is far from simple and straightforward to figure out the best set and number of genes in order to maximize information and minimize noise (Kurian et al., 2014). I must also note that this study also suggests that more than half of the cancer occurrences were not related to any of the pathogenic variants identified in this study. Therefore, the question why those seven sisters developed breast cancer without carrying the relevant *BRCA* alleles is not puzzling at all.

Another recent study aimed at assessing the psychological, behavioral, and clinical effects of the use of DTCG tests and risk estimation with the Navigenics Health Compass, a commercially available test. Participants who purchased the Health Compass were recruited, and 2,037 out of an initial cohort of 3,639 people reported whether or not they experienced changes in symptoms of anxiety, intake of dietary fat, exercise behavior, test-related distress, and the use of health-screening tests at approximately six months after testing. It was concluded that undergoing genetic testing did not result in any measurable short-term changes in psychological health, diet, exercise behavior, or use of screening tests (Bloss et al., 2011). These findings indicate that the presumed knowledge that a genetic test confers does not necessarily result in a change of attitude that might lead to disease prevention. At the same time, any sensible person would wonder: everyone accepts that a healthy diet and lifestyle and physical exercise are good. Therefore, why would one need a genetic test in order to start doing these? Even if one were not susceptible to disease these would do no harm, would they?

Nonexperts may think that genes matter and they could be a good guide to a long and healthy life. But this is not the case yet. Experts know better, and they seem to understand the current limited clinical utility of genetic testing. A poll among readers of the prestigious journal *New England Journal of Medicine* produced interesting findings. The poll presented the case of a fictional asymptomatic forty-five-year-old man who was concerned about his risk for cancer. The views of two experts were presented on whether or not they would recommend genetic screening in an asymptomatic person, and in case they did whether the whole genome or just a set of cancer-related genes should be sequenced. On the basis of this, readers of the journal were asked for their recommendations. Of the 929 readers who voted, 47 percent would recommend sequencing for cancer genes only and 12 percent would recommend sequencing the whole genome. The remaining 40 percent would not recommend any genetic testing. Nevertheless, from those readers who would recommend testing and provided comments, most expressed reservations. In particular, they questioned the appropriateness of genetic screening of any kind and suggested that additional information such as family history would be necessary for a better assessment. These results certainly reflect some concerns about genetic testing and about genome sequencing in asymptomatic persons who do not have a strong family history of diseases for which there are known genetic risk variants and available treatments (Schulte et al., 2014).

To put things in a straightforward manner: if one is going to develop a disease in "x" years (something that cannot be foretold) and live with the disease for another "y" years (how many, it is not possible to know), there are two facts independently of the values that "x" and "y" will take. The years living with the disease, designated as "y," will certainly be difficult both for the person and the family. Is there any point of making the rest of one's life, the "x" years stressful and miserable by anticipating the onset of the disease? I think not. Rather than live "x" stressful years and "y" years with the disease, you have the right not to know and live the present without worrying what the future will bring. All medicine is probabilistic; but if you are a man and you have high cholesterol, you can start taking statins and adopt a healthy diet. There is something you can do about this, in an attempt to delay heart disease. But if you are a woman and you find that you carry certain alleles such as the *BRCA1* ones, there is not much that you can really do. You could be one of the 60 out of 100 women (on average) who will develop breast cancer; but you could also be one of the 40 out of 100 women (again, on average) who will not develop breast cancer. The choice of double mastectomy that Angelina Jolie made is certainly an option; but is it the best one? This is a very personal decision that depends on other factors. If people are well informed and they have understood the advantages and the disadvantages of such decisions, then they could make their own decisions.[4]

On the basis of all these, it should be clear that genetic testing about cancer can only provide limited information. This is also currently the case for most complex diseases. What is even worse is that such concerns about the usefulness of genetic tests imply that this is the only issue, as if these are always definitive and accurate. The probabilities that a test is wrong are rarely considered, particularly if the result is the wished one. However, it is possible that a genetic test indicates that one has a disease-related variant whereas this is not the case – this is a "false positive" result. Accordingly, a genetic test may fail to identify a disease-related variant that one indeed has – this is a "false negative" result. These cases are of course rare, or at least they are less probable than an accurate test. Nevertheless, they exist. The probabilities of a "false positive" or a "false negative" result are reflected by two important characteristics of the test performed: its *sensitivity* and its *specificity*. The sensitivity of a test indicates the probability for a test to yield a true positive result. For example, if the sensitivity of a test is 99 percent, this means that 99 out of 100 tests performed will accurately indicate that these people have the respective variant; however, one of them will not, and this result will be a false positive. Now, the specificity of a test is the probability that a negative test result is a true negative one. For example, if the specificity of a test is 99 percent, this means that 99 out of 100 tests performed will accurately indicate that these people do not have the respective variant; however, one of them will have it, and this result will be a false negative one (see Annas & Elias, 2015, p. 249). The probabilities for the possible outcomes of a test are presented in Table 12.3. These are conditional probabilities, because they express the probability of an event (e.g. a test outcome being positive or negative)

---

[4] Of course, the potential may increase in the future. The first longitudinal detailed integrative personal omics profile (iPOP) of a person was published in 2012, and it was used to assess disease risk and monitor disease states for personalized medicine (Chen et al., 2012). That person was the leader of this group, who observed the onset of his type 2 diabetes. Following that, he has also published an optimistic but rather balanced and accessible guide to genomics and personalized medicine (Snyder, 2016).

TABLE 12.3 *Possible Outcomes of a Test for a DNA Variant*

Test result	DNA variant	
	Yes	No
Positive	Probability of a true positive outcome (sensitivity)	Probability of a false positive outcome
Negative	Probability of a false negative outcome	Probability of a true negative outcome (specificity)

Source: Based on Gigerenzer (2002, p. 47, Table 4-1).

given the probability of another event that has occurred (e.g. whether one has or does not have a certain DNA variant) (see Hand, 2014, p. 63).

How likely is a false result? Researchers scanned exome sequences from 478 individuals with well-studied phenotypes for potentially pathogenic variants in 17 genes related to 11 genetic disorders that are among the most medically actionable ones in adults. They also developed five variant selection algorithms with increasing sensitivity, i.e. a highly probable true positive outcome, and measured their specificity, i.e. a true negative outcome, in these seventeen genes. The results showed that variant selection algorithms with increasing sensitivity had decreased specificity. The most sensitive algorithm ranged from 88.8 to 99.6 percent specificity among the seventeen genes tested. This means that for certain genes there would be an increase in false positive results. A way to address this problem is to systematically evaluate test performance for each condition (Adams et al., 2015). This is very important to keep in mind because false positive results can cause significant distress and anxiety. A study actually attempted to document the experiences of families who had received a false-positive newborn screening result in order to improve the respective communication process. Twenty-seven parents who had received a false positive test for their children during newborn screening were interviewed. Most parents did not report long-term negative impacts of the experience, but some had experienced some worry (Schmidt et al., 2012). The communication process can certainly be improved; however, the important piece of information is for people to know before the test that a false-positive (or a false negative) result is possible. There are several reasons why genetic tests and analysis have pitfalls.

Another problem that may be easily missed can pose major problems for diagnosis. As already explained in previous chapters, several kinds of mistakes can occur either during DNA replication resulting in molecules with DNA sequences different from those of the initial ones, or during the segregation of chromosomes that takes place during cell division. These mistakes can have various outcomes such as variations in a single or a few nucleotides in the first case, whereas in the second case it is possible to have variations in the number of repeated sequences due to rearrangements in the structure of chromosomes or even variations in the number of chromosomes. Consequently, as these mistakes can take place at any stage of development, any human ends up consisting of a collection of cells that have differences in the genome that they contain. This phenomenon is called genome mosaicism. This should be kept in mind when samples are taken for DNA testing, especially in the case of prenatal testing for a disease. These analyses reflect the average genome from the cells that are examined, and especially when these are one or a few an identified abnormality might be due to the sampling of cells that differ with one another and might not exist in most cells of the organism (Lupski, 2013).

What is the conclusion from all these discussions about genes and personalized medicine? As pragmatically described in a recent book: "For the immediate future, however, we will only be able to probabilistically predict, but not prevent, most diseases" (Annas & Elias, 2015, p. 2). All that we can do at this point is compare one's estimated risk with an average population risk for a certain disease. On the one hand, being above the average population risk does not entail that one will necessarily develop a disease; therefore, one should not feel destined to have the disease even though one is currently healthy. On the other hand, being below the average population risk does not entail that one will not develop the disease; therefore, one should not feel free to smoke or be exposed to sunlight just because one's estimated genetic risk for lung or skin cancer is low. The decision is as subjective as that about whether a glass of water filled up to its middle is half-empty or half-full. The choice is yours. But does it make sense to worry about factors that we cannot control? I think not. Therefore, let's worry about we can do, e.g. adopt a healthy diet, do physical exercise, quit smoking, drink sensibly, avoid long exposure to sunlight, and stop worrying about the rest even if they are inside us. Insofar as finding a certain variant in our DNA is not "medically actionable," i.e. does not "translate into a clinically relevant action that can cure, ameliorate, or prevent a disease or other adverse impacts of health" (Annas & Elias, 2015, p. 157), genetic testing for these seems unnecessary, and perhaps pointless. This of course will probably change in the future, but for now regular medical checkups are more useful than estimating genetic risks.

Beyond these, there are broader issues that are necessary to consider at this point. These include the faults and biases in the conduct of scientific research and in the interpretation of its results, which are rarely discussed. This is, in part, a source for the public distortion of scientific findings.[5] In a highly cited article, John P. A. Ioannidis (1965–) has argued that most published research findings are false (2005). Strange as it might sound at first, this should be no surprise given that the history of science has clearly shown that most "scientific discoveries" of a certain time will be revised, or be altogether rejected at later times. What Ioannidis points out is that in many cases confirmation of research findings is not attempted by replication studies and thus conclusions are made on the basis of individual studies. This is highly problematic and is one of the reasons that we find contrasting conclusions from research on the same topic. In contrast, what is needed from researchers is that they are conscious of the possible biases in the study design and interpretation, as well as that they look at where the whole field is going and what various studies on the same question are finding. With this in mind, Ioannidis concludes with some corollaries about the probability that a research finding is indeed true:

1. *The smaller the studies conducted in a scientific field, the less likely the research findings are to be true.* Research findings are more likely to be true in scientific fields with studies involving thousands of subjects (such as the GWAS discussed earlier in this book) rather than in fields with studies involving one or a few hundred subjects.

2. *The smaller the effect sizes in a scientific field, the less likely the research findings are to be true.* Research findings are more likely to

---

[5] For an impressively accurate, extremely funny, highly informative, and a bit politically incorrect account see the "Last Week Tonight" episode "Scientific Studies" with John Oliver (www.youtube.com/watch?v=0Rnq1NpHdmw).

be true in scientific fields with large effects and higher relative risks, such as the impact of smoking on cancer or cardiovascular disease, than in scientific fields with smaller effects and relative risks, such as genetic risk factors for complex diseases.

3. *The greater the number and the lesser the selection of tested relationships in a scientific field, the less likely the research findings are to be true.* Research findings are more likely to be true in confirmatory designs or meta-analyses than in initial studies in a given field.

4. *The greater the flexibility in designs, definitions, outcomes, and analytical modes in a scientific field, the less likely the research findings are to be true.* Flexibility in all of these dimensions may increase the potential for a biased interpretation of results by interpreting what would be "negative" results as "positive."

5. *The greater the financial and other interests and prejudices in a scientific field, the less likely the research findings are to be true.* Conflicts of interest, particularly the ones that relate to funding or financial interests, and prejudice, e.g. because one is committed to a particular view or wants to defend one's personal findings, may cause a biased interpretation of results.

6. *The hotter a scientific field (with more scientific teams involved), the less likely the research findings are to be true.* When many research groups are involved in research on the same topic, it is very likely that contrasting findings will follow the initial excitement as the different teams may end up with different results because of different methods used or because they tried to confirm or refute the findings of other teams (a phenomenon Ioannidis describes as the Proteus effect: "the phenomenon of alternating extreme research claims and extremely opposite refutations").

All these point to a fact that we do not like and that we often, consciously or unconsciously, choose to ignore: that it is impossible to be 100 percent certain what the truth is in any research question. The situation could be improved by obtaining evidence from large studies or low-bias meta-analyses, but these also have their limitations. Another thing to do is to consider the totality of the evidence, i.e. the findings of all research teams and not only the statistically significant findings of one or a few teams. Finally, rather than looking for statistical significance only, researchers should carefully consider the chances that they are really testing a true rather than an untrue relationship (Ioannidis, 2005).

Another important issue is whether research findings are really useful. This is especially important for findings that may have clinical applications, as it is the case for the GWAS and other kinds of genetic studies discussed in this book. The point here is that even if all findings were true, there would still be important questions about their clinical significance and their actual usefulness in medical practice. Clinical research is useful when it can lead to a favorable change in decision making, either by itself or when integrated with other studies and evidence. There are several features that useful clinical research should have, such as:

- its problem base (Is there a health problem that is big/important enough to fix?)
- its context (Has prior evidence been systematically assessed to inform new studies?)
- its information gain (Is the proposed study large and long enough to be sufficiently informative?)
- its pragmatism (Does the research reflect real life?)
- its patient-centeredness (Does the research reflect top patient priorities?)
- its value for money (Is the research worth the money spent?)
- its feasibility (Can this research be done?)
- its transparency (Are methods, data, and analyses verifiable and unbiased?)

It has been argued that many studies, even those published in prestigious medical journals, do not satisfy all of these features; it seems that actually very few studies satisfy most or all of them. This entails that most clinical research fails to be useful not because of the findings of the various studies, but because of their design (Ioannidis, 2016).

Therefore, there also exist limitations for genomic medicine, which are not specific to it. These limitations are both of the tests themselves and of the conclusions that can be drawn from them. But these are not discussed as extensively as they should be in the public sphere. Instead, nonexperts are exposed to representations of DNA testing as infallible. This is nowhere more emphasized than in forensic DNA testing, which is often presented to provide conclusive evidence without any doubt. Perhaps the most influential means for the propagation of this view are popular TV series such as *CSI: Crime Scene Investigation.* The program's hero, Gil Grissom, made several statements that indicated a specific view of forensic evidence: "We're crime scene analysts. We're trained to ignore verbal accounts and rely instead on the evidence a scene sets before us." ... "I tend not to believe people. People lie. The evidence doesn't lie." (quoted in Lynch et al., 2008, p. ix). The message here is that, whereas people may not tell the truth, either intentionally for some personal reason or unintentionally because they may have forgotten the details, physical evidence points on its own to the truth. As already mentioned in Chapter 5, a detailed study of this TV show concluded that the dominant message is that DNA testing is common, swift, reliable, and instrumental in solving cases. It was also found that the show presents DNA analyses as quick and easy, typically taking no more than a day to complete, whereas this process may actually take days, weeks, or months (Ley et al., 2012). In other words, the usual presentation of forensic DNA analyses in TV series has contributed to the myth of their infallibility. Thus, some very important sources of error are overlooked, whereas the actual and possible pitfalls of these methods are rarely discussed. However, these pitfalls are not only more common than usually assumed, but they have also formed the basis for mistaken court decisions (for various actual cases, see Lynch et al., 2008; Krimsky & Simoncelli, 2011; Thompson, 2013).

The technique used for many years in forensic DNA analysis is known as DNA fingerprinting, and it was invented by Alec Jeffreys (1950–) in the 1980s (Jeffreys, Wilson & Thein, 1985). Its application to forensic investigations was immediately suggested (Gill et al., 1985; Jeffreys, Brookefield, & Semeonoff, 1985). This is different from the genomic analyses that I have described so far in this book that have taken place in HGP and GWAS. The first difference is that for DNA fingerprinting it makes no difference where genes are, in other words, whether a DNA sequence encodes some information or not. The second difference is that DNA fingerprinting does not examine the whole genome of an individual, but certain highly variable DNA sites, i.e. DNA sequences that vary a lot among individuals. Gene sequences tend to be quite similar among individuals as they are selectively maintained through the survival of their possessors. In contrast, sequences outside genes that are nonfunctional tend to accumulate mutations and thus vary significantly among individuals. These sequences usually differ either between their exact order of nucleotides or in the number of times that a certain DNA segment is repeated. The exact differences can be found and because of their high variability they can be used as the means to distinguish among individuals.

Let us see how this can be done. I must note that the DNA finger-printing that I am using here as an example is not the method currently used. However, it is simple enough to make my point, and I also think that more people will be familiar with it through films or textbooks.[6] All DNA analyses depend on the extraction of DNA from cells or tissues (e.g. blood, saliva, hair, semen). If this is achieved, then DNA can be cut into fragments by specific enzymes, called restriction endonucleases because they cut within the DNA in restricted sites. DNA is not cut in the same way for each individual, as these enzymes "recognize" specific sequences. If these sequences are highly variable, i.e. differ among different individuals, then the DNA of each individual will be cut in different ways. The resulting fragments will differ in size and can be separated through electrophoresis. What happens is that under the influence of an electric field, the negatively charged DNA molecules move toward the positive electrode within a porous gel. Thus, the various fragments are separated because the shorter ones move faster, whereas the longer fragments move more slowly inside the porous gel. The distribution of these fragments in the gel can then become visible. The DNA fragments are denatured to give single strands. The addition of a radioactive probe, a short DNA sequence that is complementary to the sequences of interest and that links to these, sends a signal that can be captured on X-ray film (the whole procedure is a bit complicated and is called Southern blot). Thus, a pattern of fragments, a profile that resembles a barcode, is produced for each individual (Figure 12.2a). Each profile is considered to be unique for each individual, and thus comparisons are possible.

[6] Of the methods currently used in forensic analysis, one produces the electropherogram. This is the output of a DNA analyzer showing the alleles for each individual, and it is based on a standard set of short tandem repeats (STRs) in human genomes (see Krimsky & Simoncelli, 2011, pp. 21–24; Jobling & Gill, 2004). A more recent advancement is the IrisPLEX assay, which requires only six genes to differentiate among forty shades of blue or brown eye color (Kayser & deKnijff, 2011). But, of course, even these more advanced methods are not infallible.

(a)

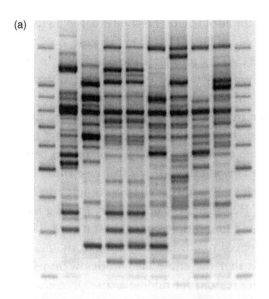

(b)

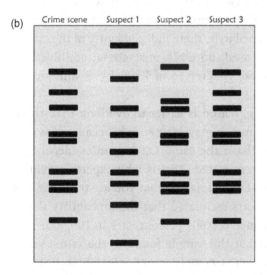

FIGURE 12.2 (a) DNA fingerprinting techniques used to provide DNA profiles like these, which scientists could then compare to one another to find matches (© Zmeel Photography). (b) On the basis of the results of DNA fingerprinting such as the ones presented in this figure, one can only infer (always to a certain degree of certainty) that the DNA from the crime scene does not come from suspects 1 and 2, as well as that it might come from suspect 3. However, in no way do these results alone suggest that suspect 3 is the criminal. Note that the actual results in (a) do not look as ideal as those in (b). Therefore, whether or not two profiles are similar is a matter of interpretation.

Let us consider a simple example of where this kind of DNA analysis can go wrong, which is presented in Figure 12.2b. In this case, only one of the samples of the suspects seems to match the one found in the crime scene, and thus may seem easy to people to infer that suspect 3 is the criminal. However, the only inference that can be made from this image is that the DNA found in the crime scene does not come from suspects 1 and 2. It cannot be concluded with absolute certainty that suspect 3 committed the crime for various reasons. One reason could be a laboratory error that cannot be excluded, e.g. because there was a cross-contamination of samples, or a mislabeling of samples. In one famous case, the Phantom of Heilbronn, the police were looking for a criminal whose DNA was found in multiple crime sites. It turned out eventually that the DNA belonged to a woman working in the manufacturing of the cotton swabs that the crime laboratories were using to collect DNA samples. Other reasons for finding a true DNA match could be that a person's biological material was found in the scene because someone, intentionally or unintentionally, transferred it there, or because that person happened to be there independently of the crime. No matter what method is used, no DNA analysis is infallible (see Gigerenzer, 2002, pp. 164–166; chapter 16 of Krimsky & Simoncelli, 2011; Thompson, 2013).

A fourth, more compelling, reason is easier to overlook than those just mentioned: a true DNA match may exist simply because it is possible for two individuals to have the same DNA profiles, depending on which DNA sequences were tested. This is why experts provide a probability for a true match. Let us imagine that you are told that in a famous murder case the experts estimated that the probability that a DNA match occurred by chance is 0.001 percent, or 1 in 100,000. This means that the probability that the sample found in the crime scene actually comes from the suspect is 99.999 percent or 99,999 in 100,000. "So they got the criminal," you may think. "It is almost certain that this is the person who committed the crime, as the probability is so high." Perhaps; perhaps not. The statement that "the probability that a DNA match occurred by chance is 0.001 percent or 1 in 100,000" could be rephrased as "one in 100,000 people is expected to show a match by chance." Now, although 1 in 100,000 might seem to be an absolute number, it is not; it depends on the context. If someone lives in a town of 10,000 people, given the mentioned probability it is highly unlikely that someone else happens by chance to have the same DNA profile. But this would be possible in a big city of 1,000,000 people, where 9 other people with the same DNA profile would exist. Therefore, the probabilities alone are not very informative; information about the context – in this case about how many people live in the area – is also important to have (Gigerenzer, 2002, pp. 6–7). This means that with a result of DNA fingerprinting such as the one presented in Figure 12.2, one can only exclude (always to a certain degree of certainty) a suspect and not confirm that one is the suspect. In particular, on the basis of this result, it can only be inferred that the DNA found in the crime scene does not come from suspects 1 and 2, and that it could come from suspect 3. However, it is impossible to be certain about the latter from the DNA evidence alone, and so additional evidence is required. We should also remember that the evidence does not say anything, but it is always interpreted by humans.

People look to genes for certainty. In this chapter I have outlined how DNA data can be used to estimate probabilistic outcomes and not to predict, as well as to point to a certain conclusion but not without possible pitfalls. There are two main reasons for this. The first are the current practical limitations. This may be overcome in the future, of course. In fact, if one considers the research tools that Morgan and his collaborators had 100 years ago, today's methods and techniques would seem at least unimaginable back then. Therefore, whatever seems to be science fiction today could become everyday practice in the future. But we are not there yet. Therefore, we need to be aware of current abilities and make sure not to distort them. This means that we should neither exaggerate what can be done, nor downplay it. Until the moment that these lines are written, we can sequence genomes and we can find associations between DNA variants and characters or disease. Nothing more. The validity of the respective methods and approaches are continuously increasing, but their utility is still limited. Fifteen years after completing the "script" of the human genome we are still struggling to decode it. We have learned to read it, but not yet to fully interpret it. But it seems that we are on the right track.

The second reason is a limitation inherent in human nature. We look for certainty in a probabilistic world, where combinations of critical, unpredictable events shape outcomes. We will never find it! We need to realize and accept that data becomes evidence only in the light of human inferential abilities, and so any conclusion depends on them. DNA tests tell us nothing about the prospect to develop a disease or about who committed a crime. Our DNA analysis – which is not infallible – can produce data from which we can draw inferences – which, again, are not infallible. Our efficiency in doing DNA analyses and in drawing inferences from DNA data are continuously improving. But there will always be limitations. This is why researchers and scientists who communicate research and its findings to society should be very careful about what language they use. This language should reflect both the actual potential and the actual limitations. Hype and optimism are inherent in human nature, and may be enhanced by financial, or other, interests. However, those who have done studies in the life sciences and who are supposed to better understand how things are than those who have no such expertise should be sensible in the claims they make and in the messages they convey about the prospects of research in genetics. This provides the topic for the closure of the book.

Concluding Remarks: How to Think and Talk about Genes?

We now live in the postgenomic era: the period after the completion of the sequencing of the human genome in which whole genome technologies became prevalent (Stevens & Richardson, 2015, p. 3). These technologies and the research in which they have been applied has brought about a total reconceptualization of genes, genetics, and genomics. This has resulted in a shift from the conception of a pregenomic genome as a static collection of distinct genes separated by "junk" DNA to the conception of a postgenomic genome as a dynamic and reactive system that regulates the expression of genes and the production of functional RNAs or proteins (Keller, 2015, p. 10). In this book I have discussed the conceptual issues one should deal with in trying to make sense of genes (and genomes). This book concludes that it is important to understand the complexity of all these phenomena in order to realize that genes are not our essences, that they do not determine who we are, and that they cannot alone explain characters and diseases.

However, research in how people perceive the impact of genes on their life shows that the idea of genes as essences that determine who and how we are is very intuitive (see Cheung et al., 2014, for an overview). For instance, three studies were conducted in order to examine what consequences people draw from a perceived genetic etiology for obesity. In the first study, 131 undergraduate students indicated whether or not they believed that obese people can control their weight, as well as whether obesity originates from a genetic disposition or environmental causes. The results suggested an association between a belief in genetic etiology for obesity and a belief that obese people cannot control their weight. These associations were further explored in a second study, in which 143 undergraduate students were asked to express their beliefs about an obesity-related phenomenon (metabolic rate) in the light of particular explanations. The results indicated that a genetic attribution for high metabolic rate was interpreted as more important than an experiential attribution. Finally, in a third study, 162 undergraduate students read one of three fictional media reports presenting a genetic explanation, a psychosocial explanation, and no explanation for obesity. The results indicated that participants who read the genetic explanation ate significantly more on a follow-up task (Dar-Nimrod et al., 2014). Such findings clearly indicate the perceived power of genes.[1]

Why do we consider genes to be so important? For some reason, we seem to think that there is something inside us that determines our essence: who we are. Apparently, if organisms have essences they must be at some deep level – this is by definition what an essence is about. But as I have explained so far, genes cannot be these essences. In contrast, we can think of the developmental capacities of organisms as their essences. The capacity to develop on the basis of a particular genetic material expressed under particular environmental conditions, exhibiting both robustness and plasticity, could be perceived as the essence of organisms (see Walsh, 2006, pp. 444–445). It is therefore time to reconsider our conception of how we come to be as we are, and to think of developmental processes as the important factors in this process instead of genes alone. As explained in Chapters 9–11 of this book, genes are implicated in development and account for differences in characters, but depend on their cellular context and a lot more that exists within our genome. It is therefore time to reconsider the way we talk about genes, and especially our metaphors for describing what we know about genes.

It is very common for scientists to use metaphors in order to describe and explain natural phenomena, processes, or mechanisms. These can become more comprehensible for nonexperts when they think about them in terms of phenomena, processes, or mechanisms they are familiar with. For example, physicists talk about "waves," although there is no medium in which they move, and "particles," even though these are not really solid. However, the big problem is that it is very easy to confuse the metaphor with the actual phenomenon. For example, saying that organisms can be thought *as if they were* machines is very different from considering them *as actually being* machines. Worse than that, the focus may eventually be on those elements of the actual phenomenon in which the metaphor is better illustrated. As a result, people may overlook other, perhaps important, aspects that do not fit well in

---

[1] There are also exist interesting, book-length accounts of the personal stories, struggles, and decisions of people dealing with genetic disease (see Klitzman, 2012; Lipkin & Luoma, 2016).

the metaphor. There are at least three reasons why it is inappropriate to attempt an analysis of organisms with a machine model in mind. First, there is no single best way to divide an organism into organs, as there are several kinds of interactions and interdependencies among them. As a result, the final phenotype depends not only on genes within the organs, but also on all kinds of interactions among cells during development. Second, organic processes have a historical contingency that makes universal explanations impossible. Any two organisms with identical genotype may have differences because of local interactions among cells during development (a phenomenon described as developmental noise). Finally, the distinction between causes and effects sometimes becomes difficult, as several systems are, for instance, self-regulated (Lewontin, 2000, pp. 3–4, 75–76).

Metaphors can be really useful, and they have indeed been useful in genetics research. As John Avise has nicely put it: "Evocative metaphors can distill an ocean of information, whet the imagination, and suggest promising channels for navigating uncharted genetic waters." This has certainly been the case for the metaphor of "information" encoded in DNA that paved the way for deciphering the genetic code, or the metaphor of the genome as the "book of life" on which the human genome project was based. However, metaphors can also mislead. Genes can actually be considered as an exemplar case of the negative impact that the bad use of metaphors and the use of bad metaphors can have. Two metaphors have been mainly used by biologists and nonbiologists in order to account for what genes "do": the genome has been described as a book and as a genetic program. However, these metaphors may have blinded biologists to genomic imperfections (see Morange, 2002, pp. 22–24; Avise, 2001, p. 86). For instance, the announcement of the first sequence of the human genome in 2000 as the outcome of the Human Genome Project was presented in BBC under the title "Reading the book of life," actually using both of these metaphors in the first lines: "The blueprint of humanity, the book of life, the software for existence – whatever you call it, decoding the entire three billion letters of human DNA is a monumental achievement."[2] The genome was thus perceived both as a book containing information and as a program through which this information is used. In a similar manner, the ENCODE (Encyclopedia

---

[2] http://news.bbc.co.uk/2/hi/indepth/scitech/2000/humangenome/760893.stm

of DNA Elements) project in 2012 was presented by CNN under the title "DNA project interprets 'book of life,'" in which it was stated that: "When the Human Genome Project sequenced the human genome in 2003, it established the order of the 3 billion letters in the genome, which can be thought of as 'the book of life'."[3] It must be noted at this point that this metaphor is inherently adopted in the notion of the "encyclopedia" employed by the researchers of the ENCODE project themselves. Regarding the program metaphor, Craig Venter, a leading figure in the sequencing of the human genome, described life as "a DNA software system" that both creates and directs the (more visible) hardware of life, such as proteins and cells.[4] Several other metaphors have been employed: DNA analysis has been described as "reading"; DNA replication has been described as "copying"; RNA synthesis has been described as "transcription"; protein synthesis has been described as "translation"; RNA modification has been described as "editing"; and more. Such metaphors are not inherently wrong and can actually help us make sense of the respective phenomena. But, nonexperts especially should always keep in mind that metaphors are a means of representation, nothing more. "Books," "software," "reading," "writing," and so on are all human inventions and thus have an inherent dimension of anthropomorphism. This needs to be made explicit, or we should otherwise avoid any unnecessary use of expressions of this kind.

However, this is not always the case. For example, the authors of a recent and well-informed book made the choice to use anthropomorphism in their descriptions of what genes do. In a section titled "Speaking of Genes," which should have been in the very beginning on the book, the authors write:

> In this book, we often write about genes as though they had intentions, as though they had consciousness. They don't, of course. Genes are nothing but stretches of DNA, complex assemblies of atoms. But when we examine the genes' properties and their consequences, it appears as if the genes were acting to ensure their own survival. This is because the evolution of genes, like all living things, is driven by the logical necessity of natural selection. For example, when we write "a cancer gene aims to secure an unfair advantage," we are using

[3] http://edition.cnn.com/2012/09/05/health/encode-human-genome/
[4] www.theguardian.com/science/2013/oct/13/craig-ventner-mars

shorthand for "mutations to an oncogene that cause an increased growth rate of the cells that carry the mutation will over time lead to an increase in the total fraction of body cells that carry the mutation." Anthropomorphizing provides a convenient shorthand for discussing many processes; while this helps to develop an intuition about natural selection, we need to remember the full description behind the shorthand. (Yanai & Lercher, 2016, pp. 37–38)

Indeed, we need! But will all people be able to do this? These authors are of course aware that genes do not have intentions. But is this equally clear to their intended audience? I suggest that experts should be very careful in using expressions like these, especially when talking about genes that are easily perceived by people as all-powerful and distort the messages that should be conveyed (see Sullivan, 1995, for a critique of Dawkins' metaphor that inspired the authors of the aforementioned book; see also Sober & Lewontin, 1982).

My suggestion therefore is for scientists to always be explicit about the limits of the metaphors they use. We can say that genes "encode" some "functional" products, insofar as we clearly explain that this is just a way of representing the informational properties of DNA that are not inherent, and that make sense only in the cellular context in which they can in turn be used as a resource for the production of molecules that contribute to the maintenance and the roles of self-regulated, living systems. We should also explain that often metaphors are used because we ignore the details and so they have a heuristic value both in explaining the respective phenomena and in guiding further research. It seems that we will have to rely on metaphors, and that for various reasons they cannot be avoided. Therefore, what we need to do is use metaphors that are more inclusive and represent more accurately the respective phenomena. For example, we need to stop thinking in terms of genes only and start thinking in terms of genomes (or genetic material) that include genes and various other sequences. We also need to replace the concept of gene action with that of gene interaction. This means that we should refrain from talking about genes that do this or that, and refer to genes that interact with other genes and with their environment. With simple changes like these we at least can give a better sense of the complexity of these phenomena.

But this is not enough, if scientists are not actively engaged in genetics education and outreach, and if they do not have the required preparation for that. During the early stages of development and application of new technologies, there is not sufficient experience to foresee how these might be used, which ethical and societal problems might occur, who might object to their use and why, and how it would be possible to deal with the problems that might occur. Therefore, it is necessary to communicate unbiased, well-informed, and well-argued estimates of potential questions and problems, as well as potential answers to all these. For example, researchers who publish first their findings of an association between a gene and a condition should make clear in their publications and in other communications why their findings are not ready for commercial applications. In my view, researchers have a professional and social responsibility to ensure that the public communication and discussion of topics and questions related to their research are explicit about their actual application and about their limitations. Otherwise, those who can benefit from the respective applications, or those who fear them, could take over – in an irresponsible manner – the role of informing the nonexperts. Society needs to have a robust understanding and develop a critical stance before making any decisions. New research and technologies can neither be accepted nor rejected uncritically (Murray et al., 2010; Reydon et al., 2012).

However, even if most scientists would agree that they should be engaged in genetics communication and outreach, particular barriers to their participation in these kinds of activities seem to exist, as a study involving twenty academic geneticists in the United States has shown. In particular, these barriers include: (1) the lack of time to engage in activities besides those required to earn tenure; (2) the receipt of no credit toward tenure for public engagement activities; (3) the fear of professional stigmatization by being too visible in the public sphere; (4) the unawareness of opportunities for public engagement and participation in policy formation; and (5) the lack of necessary communications skills (Mathews et al., 2005). Although these views were expressed by a small number of geneticists and so generalizations are difficult to make, these barriers seem likely to exist in many cases. Therefore, perhaps universities should reconsider their criteria for tenured professorships and evaluate candidates not only in terms of their publishing records and research agendas, but also in terms of their participation in outreach activities and science communication. Under the appropriate institutional reforms, researchers could have the opportunities to develop the necessary skills for this purpose. These skills include the ability of researchers to communicate their research, and the relevant results and conclusions, to journalists who in turn communicate these to the public.

Some concrete suggestions that might facilitate this communication and make it accurate have also been made. The first suggestion is that instead of just chatting with the journalist, scientists should prepare the message they want to convey by writing the key points in lay language and testing its readability. The second suggestion is for scientists to try to avoid hype stemming from their enthusiasm for their own work and the demands for funding. Thus, instead of making ambitious predictions for future applications, they should describe the technical and social roadblocks and explain why current findings could be overturned. A third suggestion is to avoid gene-centered talk, such as reference to "genes for," and in general genetic determinism that seems so intuitive to nonexperts. As two organisms with the same genes might exhibit different phenotypes in different environments, scientists should make clear that it is the interaction of genes and environment that determine phenotypic outcomes. Finally, the fourth suggestion is to avoid labels that enhance discrimination among different human groups or populations, such as intuitive notions of "race" or those that refer to people carrying an allele that is related to a specific disease. Instead, scientists should try to carefully and precisely characterize human groups or populations (Condit, 2007). Overall, these suggestions imply that scientists should carefully prepare any public communication of their work. This is very important as what is qualified as news by journalists and scientists might be different. Journalists may think of well-established work as "old news"; they may be more interested in publishing new findings that may be tentative and provisional for scientists (Nelkin, 1995, pp. 164–165).

Jon Beckwith (1935–), who addressed the XYY myth in the 1970s (Beckwith & King, 1974), has recently described cases of distortion of scientific findings and of subsequent misinformation. On the basis of these, he made three very important suggestions that should be given careful consideration: (1) that a broader education of scientists is necessary so that they become aware of how scientific findings have been used during history and understand the connections between a scientist's work and its social impact; (2) that editors and reviewers of scientific journals ensure that the conclusions made by authors of scientific articles are not exaggerated and do not include speculations with potential harmful social implications; and (3) that scientists consider as an important responsibility not only to be careful in how they present their own work, but also to comment on and provide alternative views to reports that include exaggerations and unnecessary speculations (Beckwith, 2013). It is therefore necessary that scientists are trained in a way that helps them become aware of controversies on socioscientific issues, and that they are motivated to participate in science communication.[5]

Some philosophical reflections in undergraduate and graduate biology courses could also be very useful. There are important conceptual issues to be resolved, including the validity of simple "Mendelian" inheritance that may be presented as the norm and that ignores the complexities of development (Jamieson & Radick, 2013); the conception of "genes for" characters despite the various kinds of interactions among gene products implicated in the development of characters (Burian & Kampourakis, 2013); the misconceived notion of "Nature vs. Nurture" that overlooks that both genetic and nongenetic components are essential for biological processes (Moore, 2013b); or the interaction between genetics research and the wider social context, in particular the commercialized social context within which research in genomics is carried out that raises concerns about whether the objectivity of scientists is compromised (Gannett, 2013). These are just some examples. I hope that in this book I have shown the importance of philosophical reflection about concepts. The meanings of concepts should not be taken for granted. Rather, these should be continuously refined and clarified – both among experts and within science education and communication.

To summarize, scientists need to be aware of how nonexperts think about genes; to use appropriate metaphors and to be explicit about their limitations; to be aware of the social impact of their work and comment on that; and to actively participate in education and outreach activities. These in turn require a broader culture and studies than what undergraduate studies usually provide. Concepts and conceptual issues should be given more attention in education and outreach. This would be the first step for an effective "translation" of these concepts from the

---

[5] For this purpose, Bruno J. Strasser and I are currently teaching a compulsory course, titled Biology and Society, to the second year undergraduate biology students at the Section of Biology of the University of Geneva.

scientific context to the public sphere. Metaphors will always be there, and we can make an appropriate use of good metaphors that will help nonexperts make sense of genes, and more broadly of genetics research. This is a much needed step before anyone attempts to understand the impact of this research for one's life.

A final important element is what, in science education, is described as nature of science. Understanding the concepts and content knowledge of genetics is certainly important. However, it is also important for nonexperts to understand some important aspects of the nature of scientific knowledge and of the nature of the processes of scientific inquiry through which this knowledge is produced (Kampourakis, 2016). These relate to the various limitations of science discussed in Chapter 12. If nonexperts hold conceptions of science as certain and infallible, then it is natural for them to have unreasonable expectations from genetics research. But science is uncertain and fallible, and genetics is no exception. In a sense, the nature of science is counterintuitive and perhaps unnatural for nonexperts, as one author put it (Wolpert, 1992). Stuart Firenstein has elegantly shown that ignorance and failure are essential aspects of science. On the one hand, it is ignorance, not knowledge and certainty, that drives science because it motivates scientists to look for answers (Firenstein, 2012). On the other hand, it is failure that makes science so successful because it is an integral part of the process for advancing and eventually succeeding (Firenstein, 2016). Another important distinction to keep in mind is that between the current state of research and where we want to go. Actual potential is different from wishful thinking and we should keep that in mind. The advancements in genetics research (as in many other domains) during the past 100 years are astonishing, and we may develop unthinkable skills in manipulating genes before long. But we should wait until this happens and not rush to (intentionally or unintentionally) misinform nonexperts about the actual potential.

At the same time, although genetic explanations are discussed in schools, media, and elsewhere, this is not always done with some explicit reflection on the inquiry processes and the scientific knowledge that underlie these explanations, as well as on the respective scientific theories and models from which these explanations are derived. One important missing element from these discussions is that much of what is presented in results and conclusions, for example, in many of the figures of this book, are themselves scientific models or are somehow related to such models. The problem is that the properties of these models are not usually discussed in an explicit manner. For example, the double helix of DNA (Figure 3.2), which has indeed become a cultural icon (Nelkin & Lindee, 2004), is more often than not presented as the true structure of the molecule. It is rarely explained that this is a scientific model, which only exhibits similarities with the actual molecule of DNA, and which does not portray how the molecule of DNA actually is. Scientific models are representational tools with specific features: they exhibit similarities with aspects of real systems; they do not represent whole systems, neither are they true of them; they are neither entirely precise, nor entirely accurate; and they are used by scientists for a particular purpose in a specific context. Maps are a nice example that can help nonexperts to keep in mind the properties of scientific models, even if it is not the best analogue for all models (Giere, 2006, pp. 72–80). Insofar as nonexperts understand the features of the structural and functional models used in genetics, they might better understand what to expect and what not to expect from these models.

So, how should we think about genes? Here is the take-home message of this book: Genes were initially conceived as immaterial factors with heuristic value for research, but along the way they acquired a parallel material identity as DNA segments. The two identities never converged completely, and therefore the best we can do so far is to think of genes as DNA segments that encode functional products. There are neither "genes for" characters nor "genes for" diseases. Genes do nothing on their own, but are important resources for our self-regulated organism. If we insist in asking what genes do, we can accept that they are implicated in the development of characters and disease, and that they account for variation in characters in particular populations. Beyond that, we should remember that genes are part of an interactive genome that we have just begun to understand, the study of which currently has various limitations. Genes are not our essences, they do not determine who we are, and they are not the explanation of who we are and what we do. Therefore, we are not prisoners of any genetic fate. This is what the present book has aimed to explain.

If you agree with this and if you are now able to explain this to others and help them overcome the intuitive conception of "genes for," I think you should be confident that you can make sense of genes.

Further Reading

This section aims to provide suggestions for books that one might read in order to explore further the topics discussed in the present book. The list that follows is in no way exhaustive. For simplicity, only the main titles of the books are mentioned here; full details are provided in the list of references, where you will find a lot more books and articles that might be interesting.

If you want to read a single book in order to find out more about the distortions and myths about genes, as well as about the social implications of research in genetics, this should be *Genetic Explanations*, edited by Sheldon Krimsky and Jeremy Gruber. An older collection of this kind is *The Double-Edged Helix* edited by Joseph S. Alper, Catherine Ard, Adrienne Asch, Jon Beckwith, Peter Conrad, and Lisa N. Geller. Another relatively old but still very valuable account of the social implications of research in genetics is given in *The Limits and Lies of Human Genetic Research* by Jonathan Kaplan.

Several books have explicitly addressed, and argued against, genetic determinism. These include *The DNA Mystique* by Dorothy Nelkin and Susan Lindee; *Exploding the Gene Myth* by Ruth Hubbard and Elijah Wald; *The Doctrine of DNA* and *The Triple Helix* by Richard Lewontin; *The Dependent Gene* by David S. Moore; *The Misunderstood Gene* by Michel Morange; *The Mirage of a Space between Nature and Nurture* by Evelyn Fox Keller; and *Genes, Cells and Brains* by Hilary Rose and Stephen Rose. For the relation between nature and nurture, Evelyn Fox Keller's book *The Mirage of a Space between Nature and Nurture* is the best one to start with. James Tabery's *Beyond Versus* provides an interesting account of relevant historical controversies on this topic.

For an overview of the history of the gene concept, one could start with Evelyn Fox Keller's *The Century of the Gene*. A very interesting book is *The Gene: From Genetics to Postgenomics* by Hans-Jörg Rheinberger and Staffan Müller-Wille, which presents a rich and concise account of the role that the gene concept has played in research. Another useful resource is the "Gene" entry by Hans-Jörg Rheinberger, Staffan Müller-Wille, and Robert Meunier in *Stanford Encyclopedia of Philosophy*. A more detailed account of the history of the gene concept is given in the book *The Concept of the Gene in Development and Evolution*, edited by Peter Beurton, Raphael Falk, and Hans-Jörg Rheinberger. For a philosophical account of genetics, a very good book is *Genetics and Philosophy* by Paul Griffiths and Karola Stotz. Another interesting book of this kind is *What Genes Can't Do* by Lenny Moss. For philosophical accounts focusing on genomics research, the books *Genomes and What to Make of Them* by Barry Barnes and John Dupré, and *Postgenomics*, edited by Hallam Stevens and Sarah Richardson, will be rewarding readings.

The first histories of genetics were written by two researchers of the era of classical genetics. These are Alfred Sturtevant's *A History of Genetics* and Leslie Dunn's *A Short History of Genetics*. They were both published in 1965, on the centenary of the presentation of Mendel's paper, and they are good places to start. Another overview of twentieth-century genetics is provided in *Genetic Analysis* by Raphael Falk, whereas a comprehensive overview is also given in the respective chapters of Jan Sapp's *Genesis*. A very important resource that presents the study of heredity in its historical and cultural contexts is *A Cultural History of Heredity* by Staffan Müller-Wille and Hans-Jörg Rheinberger. Many classic books and papers of genetics are made available from Electronic Scholarly Publishing at www.esp.org. The journal *Genetics* also contains numerous "perspective" articles on the history of genetics.

There also exist excellent books for the various historical periods. For the pre-1900 period and the origins of mendelian genetics, the best book is definitely Robert Olby's *Origins of Mendelism*. Another interesting book on this topic is *The Mendelian Revolution* by Peter Bowler. For the life and work of Mendel the most authoritative source are Vitezlav Orel's writings, such as his book *Mendel*. A complete overview of classical genetics is given in Elof Axel Carlson's *Mendel's Legacy*. A detailed study of the work done by Morgan and his colleagues is Robert Kohler's book *Lords of the Fly*. For the work of Morgan, one might also look at Garland Allen's *Thomas Hunt Morgan: the Man and His Science*. The single best resource on eugenics is the book *In the Name of Eugenics* by Daniel Kevles. Diane Paul's *The Politics of Heredity* and *Controlling Human Heredity* are also very useful resources. For the history of the emergence of genetic medicine from eugenics, two interesting books are *Moments of Truth in Genetic Medicine* by Susan Lindee, and *The Science of Human Perfection* by Nathaniel Comfort.

Coming to molecular biology, an amazing resource is the *Eighth Day of Creation* by Horace Freeland Judson, who wrote it based on interviews with most of the scientists involved. Michel Morange also gives a wide-ranging overview of the history of molecular biology in his likewise titled book. For the background and the events leading to the model of the double helix of DNA, one should read Robert Olby's *The Path to the Double Helix*. For a detailed account of how the genetic code was solved, one should look at Lily Kay's *Who Wrote the Book of Life*, and Matthew Cobb's *Life's Greatest Secret*. Finally, detailed and interesting narratives of the human genome project and of the development of companies producing genetic tests and DNA-sequencing machines are given in *Cracking the Genome* and in *The $1000 Genome*, respectively, both authored by Kevin Davies.

Several recent books provide accounts of current research in genomics. Kat Arney's *Herding Hemingway's Cats* provides an interesting account of recent findings about genes through interviews with several leading researchers. An accessible introduction to genetics that can also serve as a primer for the basics in molecular genetics is *Genetic Twists of Fate* by Stanley Fields and Mark Johnston. John Parrington's *The Deeper Genome* is a very informative and very updated account of genomics research so far. Another useful resource on this topic is Nessa Carey's *Junk DNA*.

Epigenetics is a recent and difficult-to-treat topic, because it differs from conventional textbook biology and because there is still a lot that we do not know. A nice, readable introduction is *Epigenetics* by Richard Francis. However, if you really want to understand the details without getting lost in them, the best book to start with is *The Developing Genome* by David Moore. Another interesting and detailed, but a bit more technical account, is given by Nessa Carey in *The Epigenetics Revolution*. For a broader discussion of nongenetic inheritance under an evolutionary perspective, an excellent book is *Evolution in Four Dimensions* by Eva Jablonka and Marion Lamb. For an introduction to evolutionary theory itself, and to the difficulties in understanding it, my book *Understanding Evolution* might be useful.

Genetics makes more sense in the light of development, and so it is absolutely necessary that one understands what development is and how it takes place. A very nice starting point is the book *Life Unfolding* by Jamie Davies, which provides a readable and accurate narrative of human development. A similarly readable account of human development and other aspects of human life is given in *How We Live and Why We Die* by Lewis Wolpert. *Developmental Biology* by the same author also provides a useful and brief account of various aspects of development in various organisms. A wonderful collection of philosophical accounts for genes in development and against gene centrism is *Genes in Development* edited by Eva Neumann-Held and Christoph Rehmann-Sutter.

There are many books about how to best handle the information the genetic and genomic tests might provide. *Genomic Messages* by George

Annas and Sherman Elias provides both an updated account of the current issues at stake and sensible suggestions on how to deal with them. Other books are written by physicians and provide accounts of the struggles and questions that people have to face when making actual decisions about genetic testing. Two such books are *Am I My Genes?* by Robert Klitzman and *The Age of Genomes* by Steven Lipkin and Jon Luoma. An optimistic, but rather balanced and accessible, guide to genomic medicine is the book *Genomics and Personalized Medicine* by Michael Snyder.

Some broader reflection is crucial for understanding genetics research. Two excellent and highly readable books about how science is done are Stuart Firenstein's *Ignorance* and *Failure*. *Newton's Apple and Other Myths about Science*, edited by Ronald Numbers and me, includes short essays that debunk popular historical myths about science. If you want to understand probabilities and risks, Gerd Gigerenzer's *Calculated Risks* is a very readable, accessible, and enjoyable account of the relevant issues. A similarly interesting and useful introduction to probabilities is *The Improbability Principle* by David Hand.

For those interested in reading the details of the respective research and scholarship, there are numerous citations in the main text of the present book that direct the reader to the lengthy list of references at the end of this book. The citations follow each point made, so the interested reader should read the article or book cited at the end of every argument or piece of evidence. Beyond that, the literature is enormous. If you really want to start exploring journals on genetics, it is best to start with reviews on the respective topics published in *Nature Reviews Genetics, Cell, Nature* and *Science*.

References

Abifadel, M., Rabès, J. P., Jambart, S. et al. (2009). The molecular basis of familial hypercholesterolemia in Lebanon: spectrum of LDLR mutations and role of PCSK9 as a modifier gene. *Human Mutation*, 30(7), E682–E691.

Abifadel, M., Varret, M., Rabès, J.-P. et al. (2003). Mutations in PCSK9 cause autosomal dominant hypercholesterolemia. *Nature Genetics*, 34, 154–156.

Adams, M. C., Evans, J. P., Henderson, G. E., Berg, J. S., & GeneScreen Investigators. (2015). The promise and peril of genomic screening in the general population. *Genetics in Medicine*, 18(6), 593–599.

Adkins, N. L., & Georgel, P. T. (2010). MeCP2: structure and function. *Biochemistry and Cell Biology*, 89(1), 1–11.

Alberts, B., Bray, D., Hopkin, K. et al. (2010). *Essential Cell Biology* (3rd edn). New York and London: Garland Science.

Alexandrov, L. B., Jones, P. H., Wedge, D. C. et al. (2015). Clock-like mutational processes in human somatic cells. *Nature Genetics*, 47(12), 1402–1407.

Alfirevic, Z., Sundberg, K., & Brigham, S. (2009). Amniocentesis and chorionic villus sampling for prenatal diagnosis. *Cochrane Review*, 1, 1–142.

Alkuraya, F. S. (2015). Human knockout research: new horizons and opportunities. *Trends in Genetics*, 31(2), 108–115.

Allen, G. E. (1997). The social and economic origins of genetic determinism: a case history of the American eugenics movement, 1900–1940, and its lessons for today. *Genetica*, 99, 77–88.

Allen, G. E. (1978a). *Life Science in the Twentieth Century*. Cambridge: Cambridge University Press.

Allen, G. E. (1978b). *Thomas Hunt Morgan: The Man and His Science*. Princeton: Princeton University Press.

Allen, G. E. (2003). Mendel and modern genetics: the legacy for today. *Endeavour*, 27, 63–68.

Allen, G. E. (2014). Origins of the classical gene concept, 1900–1950: genetics, mechanistic philosophy, and the capitalization of agriculture. *Perspectives in Biology and Medicine*, 57(1), 8–39.

Allen, H. L., Estrada, K., Lettre, G. et al. (2010). Hundreds of variants clustered in genomic loci and biological pathways affect human height. *Nature*, 467(7317), 832–838.

American Cancer Society (2016). *Cancer Facts & Figures 2016*. Atlanta: American Cancer Society.

Angelman, H. (1965). "Puppet" children: a report on three cases. *Developmental Medicine & Child Neurology*, 7(6), 681–688.

Annas, G. J., & Elias, S. (2014). 23andMe and the FDA. *New England Journal of Medicine*, 370(11), 985–988.

Annas, G. J., & Elias, S. (2015). *Genomic Messages: How the Evolving Science of Genetics Affects Our Health, Families, and Future*. New York: HarperOne.

Arney, A. (2016). *Herding Hemingway's Cats: Understanding How Our Genes Work*. London & New York: Bloomsbury Sigma.

Arthur, W. (2004). *Biased Embryos and Evolution*. New York: Cambridge University Press.

Avery, O. T., MacLeod, C. M., & McCarty, M. (1944). Studies of the chemical nature of the substance inducing transformation of pneumococcal types: induction of transformation by a desoxyribonucleic acid fraction isolated from Pneumococcus Type III. *Journal of Experimental Medicine*, 79, 137–158.

Avise, J. C. (2001). Evolving genomic metaphors: a new look at the language of DNA. *Science*, 294(5540), 86–87.

Baltimore, D. (1970). RNA-dependent DNA polymerase in virions of RNA tumor viruses. *Nature*, 226, 1209–1211.

Baltimore, D. (1971). Expression of animal virus genomes. *Bacteriological Reviews*, 35(3), 235.

Baran, Y., Subramaniam, M., Biton, A. et al. (2015). The landscape of genomic imprinting across diverse adult human tissues. *Genome Research*, 25(7), 927–936.

Barnes, B., & Dupré, J. (2008). *Genomes and What to Make of Them*. Chicago: University of Chicago Press.

Barrell, B. G., Air, G. M., & Hutchison, C. A. 3rd (1976). Overlapping genes in bacteriophage phiX174. *Nature*, 264(5581), 34–41.

Bartolomei, M. S. (2009). Genomic imprinting: employing and avoiding epigenetic processes. *Genes & Development*, 23(18), 2124–2133.

Bateson, P., & Gluckman, P. (2011). *Plasticity, Robustness, Development and Evolution*. Cambridge: Cambridge University Press.

Bateson, W. (1902). *Mendel's Principles of Heredity: A Defence*. London: Cambridge University Press.

Bazak, L., Haviv, A., Barak, M. et al. (2014). A-to-I RNA editing occurs at over a hundred million genomic sites, located in a majority of human genes. *Genome Research*, 24(3), 365–376.

Beadle, G. W., & Tatum, E. L. (1941). Genetic control of biochemical reactions in Neurospora. *Proceedings of the National Academies of Science, USA*, 27, 499–506.

Beckwith, J. (2002). Genetics in society: society in genetics. In Alper, J., Ard, C., Asch, A. et al. (Eds.) *The Double-Edged Helix: Social Implication of Genetics in a Diverse Society*. Baltimore: Johns Hopkins University Press, 39–57.

Beckwith, J. (2013). The persistent influence of failed scientific ideas. In Krimsky, S. and Gruber J. (Eds). *Genetic Explanations: Sense and Nonsense*. Cambridge MA: Harvard University Press, 173–185.

Beckwith, J., & King, J. (1974). The XYY syndrome: a dangerous myth. *New Scientist*, 64(923), 474–476.

Benne, R., Van Den Burg, J. Brakenhoff, J. P. J. et al. (1986). Major transcript of the frameshifted coxll gene from trypanosome mitochondria contains four nucleotides that are not encoded in the DNA. *Cell*, 46, 819–826. .

Benzer, S. (1955). Fine structure of a genetic region in bacteriophage. *Proceedings of the National Academies of Science USA*, 41, 344–354.

Benzer, S. (1957). The elementary units of heredity. In *Symposium on the Chemical Basis of Heredity*. Baltimore: Johns Hopkins University Press, pp. 70–93.

Berget, S. M., Moore, C., & Sharp, P. A. (1977). Spliced segments at the 5'terminus of adenovirus 2 late mRNA. *Proceedings of the National Academy of Sciences*, 74(8), 3171–3175.

Berry, S. A., Brown, C., Grant, M. et al. (2013). Newborn screening 50 years later: access issues faced by adults with PKU. *Genetics in Medicine*, 15(8), 591–599.

Betteridge, D. J. (2013). Cardiovascular endocrinology in 2012: PCSK9 – an exciting target for reducing LDL-cholesterol levels. *Nature Reviews Endocrinology*, 9(2), 76–78.

Beurton, P., Falk, R., & Rheinberger, H. J. (Eds.) (2000). *The Concept of the Gene in Development and Evolution. Historical and Epistemological Perspectives.* Cambridge: Cambridge University Press.

Bhattacharyya, M. K., Smith, A. M., Ellis, T. N., Hedley, C., & Martin, C. (1990). The wrinkled-seed character of pea described by Mendel is caused by a transposon-like insertion in a gene encoding starch-branching enzyme. *Cell*, 60(1), 115–122.

Bianchi, D. W. (2012). From prenatal genomic diagnosis to fetal personalized medicine: progress and challenges. *Nature Medicine*, 18(7), 1041–1051.

Bickel, H., Gerrard, J., & Hickmans, E. M. (1953). Influence of phenylalanine intake on phenylketonuria. *The Lancet*, 265, 812–813.

Biggs, A., Hagins, W. C., Holliday, W. G., Kapicka, C. L., & Lundgren, L. (2009). *Glencoe Science: Biology.* New York: McGraw-Hill.

Bird, A., Taggart, M., Frommer, M., Miller, O. J., & Macleod, D. (1985). A fraction of the mouse genome that is derived from islands of nonmethylated, CpG-rich DNA. *Cell*, 40(1), 91–99.

Blau, N., Shen, N., & Carducci, C. (2014). Molecular genetics and diagnosis of phenylketonuria: state of the art. *Expert Review of Molecular Diagnostics*, 14(6), 655–671.

Blau, N., van Spronsen, F. J., & Levy, H. L. (2010). Phenylketonuria. *The Lancet*, 376(9750), 1417–1427.

Bloss, C. S., Schork, N. J., & Topol, E. J. (2011) Effect of direct-to-consumer genome-wide profiling to assess disease risk. *New England Journal of Medicine*, 364, 524–534.

Borzekowski, D. L. G., Guan, Y., Smith, K. C., Erby, L. H., & Roter, D. L. (2013). The Angelina effect: immediate reach, grasp, and impact of going public. *Genetics in Medicine*, 16, 516–521.

Botstein, D. (2015). *Decoding the Language of Genetics.* Cold Spring Harbor: Cold Spring Harbor Laboratory Press.

Bowler, P. J. (1989). *The Mendelian Revolution: The Emergence of Hereditarian Concepts in Modern Science and Society*. Baltimore: Johns Hopkins University Press.

Braam, J., & Davis, R. W. (1990). Rain-, wind-, and touch-induced expression of calmodulin and calmodulin-related genes in Arabidopsis. *Cell, 60*(3), 357–364.

Brannigan, A. (1979). The reification of Mendel. *Social Studies of Science, 9*, 423–454.

Brannigan, A. (1981). *The Social Basis of Scientific Discoveries*. Cambridge: Cambridge University Press.

Braude, P., Pickering, S., Flinter, F., & Ogilvie, C. M. (2002). Preimplantation genetic diagnosis. *Nature Reviews Genetics, 3*(12), 941–955.

Brenner, S. (1998). Refuge of spandrels. *Current Biology, 8*(19), R669.

Brenner, S., Jacob, F., & Meselson, M. (1961). An unstable intermediate carrying information from genes to ribosomes for protein synthesis. *Nature, 190*, 576–581.

Britten, R. J., & Davidson, E. H. (1969). Gene regulation for higher cells: a theory. *Science, 165*(3891), 349–357.

Brookes, A. J. (1999). The essence of SNPs. *Gene, 234*(2), 177–186.

Brooks, W. K. (1883). *The Law of Heredity: A Study of the Cause of Variation and the Origin of Living Organisms* (2nd ed.). Baltimore and New York: John Murphy & Co.

Brownell, J. E., & Allis, C. D. (1996). Special HATs for special occasions: linking histone acetylation to chromatin assembly and gene activation. *Current Opinion in Genetics & Development, 6*(2), 176–184.

Brunner, H. G., Nelen, M. R., Van Zandvoort, P. et al. (1993). X-linked borderline mental retardation with prominent behavioral disturbance: phenotype, genetic localization, and evidence for disturbed monoamine metabolism. *American Journal of Human Genetics, 52*(6), 1032–1039.

Brunner, H. G., Nelen, M., Breakefield, X. O. et al. (1993). Abnormal behavior associated with a point mutation in the structural gene for monoamine oxidase A. *Science, 262*(5133), 578–580.

Brush, S. G. (1978). Nettie M. Stevens and the discovery of sex determination by chromosomes. *Isis, 69*(2), 163–172.

Burbridge, D. (2001). Francis Galton on twins, heredity and social class. *The British Journal for the History of Science, 34*(3), 323–340.

Burian, R. M., & Kampourakis, K. (2013). Against "genes for": could an inclusive concept of genetic material effectively replace gene concepts? In K. Kampourakis (Ed.) *The Philosophy of Biology: A Companion for Educators*. Dordrecht: Springer, 597–628.

Byrd, A. L., & Manuck, S. B. (2014). MAOA, childhood maltreatment, and antisocial behavior: meta-analysis of a gene-environment interaction. *Biological Psychiatry, 75*(1), 9–17.

Carey, N. (2012). *The Epigenetics Revolution: How Modern Biology Is Rewriting Our Understanding of Genetics, Disease, and Inheritance*. New York: Columbia University Press.

Carey, N. (2015). *Junk DNA: A Journey through the Dark Matter of the Genome*. New York: Columbia University Press.

Carlson, E. A. (2004) *Mendel's Legacy: The Origin of Classical Genetics*. Cold Spring Harbor, NY: Cold Spring Harbor Laboratory Press.

Carver, R., Waldahl, R., & Breivik, J. (2008). Frame that gene. *EMBO Reports*, 9(10), 943–947.

Carver, R. B., Rødland, E. A., & Breivik, J. (2013). Quantitative frame analysis of how the gene concept is presented in tabloid and elite newspapers. *Science Communication*, 35(4), 449–475.

Casey, M. D., Segall, L. J., Street, D. R. K., & Blank, C. E. (1966). Sex chromosome abnormalities in two state hospitals for patients requiring special security. *Nature*, 209, 641–642.

Caspi, A., McClay, J., Moffitt, T. E. et al. (2002). Role of genotype in the cycle of violence in maltreated children. *Science*, 297(5582), 851–854.

Castera, J., & Clement, P. (2014). Teachers' conceptions about genetic determinism of human behaviour: a survey in 23 countries. *Science & Education*, 23(2), 417–443.

Castle, W. E. (1903). Mendel's law of heredity. *Science*, 18(456), 396–406.

Cattanach, B. M. (1986). Parental origin effects in mice. *Journal of Embryology and Experimental Morphology*, 97(Supplement), 137–150.

Cattanach, B. M., & Kirk, M. (1985). Differential activity of maternally and paternally derived chromosome regions in mice. *Nature*, 315(6019), 496–498.

Caulfield, T., & McGuire, A. L. (2012). Direct-to-consumer genetic testing: perceptions, problems, and policy responses. *Annual Review of Medicine*, 63, 23–33.

Cech, T. R., & Steitz, J. A. (2014). The noncoding RNA revolution – trashing old rules to forge new ones. *Cell*, 157(1), 77–94.

Centerwall, S. A., & Centerwall, W. R. (2000). The discovery of phenylketonuria: the story of a young couple, two retarded children, and a scientist. *Pediatrics*, 105(1), 89–103.

Chen, R., Mias, G. I., Li-Pook-Than, J. et al. (2012). Personal omics profiling reveals dynamic molecular and medical phenotypes. *Cell*, 148(6), 1293–1307.

Chen, R., Shi, L., Hakenberg, J. et al. (2016). Analysis of 589,306 genomes identifies individuals resilient to severe Mendelian childhood diseases. *Nature Biotechnology*, 34(5), 531–538.

Cheung, B., Dar-Nimrod, I., & Gonsalkorale, K. (2014). Am I my genes? Perceived genetic etiology, intrapersonal processes, and health. *Social and Personality Psychology Compass*, 8(11), 626–637.

Chow, L. T., Gelinas, R. E., Broker, T. R., & Roberts, R. J. (1977). An amazing sequence arrangement at the 5' ends of adenovirus 2 messenger RNA. *Cell*, 12(1), 1–8.

Cobb, M. (2006). Heredity before genetics: a history. *Nature Reviews Genetics*, 7, 953–958.

Cobb, M. (2015). *Life's Greatest Secret: The Story of the Race to Crack the Genetic Code*. London: Profile Books.

Cohen, J. C., Boerwinkle, E., Mosley, T. H., Jr., & Hobbs, H. H. (2006). Sequence variations in PCSK9, low LDL, and protection against coronary heart disease. *New England Journal of Medicine*, 354, 1264–1272.

Collins, F. S. (1999). Medical and societal consequences of the Human Genome Project. *New England Journal of Medicine*, 341(1), 28–37.

Collins, F. S. (2010). *The Language of Life: DNA and the Revolution in Personalized Medicine*. New York: Harper.

Collins, F. S., & Varmus, H. (2015). A new initiative on precision medicine. *New England Journal of Medicine*, 372, 793–795.

Comfort, N. (2012). *The Science of Human Perfection: How Genes Became the Heart of American Medicine*. New Haven & London: Yale University Press.

Condit, C. M. (1999). How the public understands genetics: non-deterministic and non-discriminatory interpretations of the "blueprint" metaphor. *Public Understanding of Science*, 8, 169–180.

Condit, C. M. (2007). How geneticists can help reporters to get their story right. *Nature Reviews Genetics*, 8, 815–820.

Condit, C. M. (2010a). Public attitudes and beliefs about genetics. *Annual Review of Genomics and Human Genetics*, 11, 339–359.

Condit, C. M. (2010b). Public understandings of genetics and health. *Clinical Genetics*, 77, 1–9.

Condit, C. M., Ofulue, N., & Sheedy, K. M. (1998). Determinism and mass-media portrayals of genetics. *American Journal of Human Genetics*, 62(4), 979–984.

Condit, C. M., Ferguson, A., Kassel, R. et al. (2001). An exploratory study of the impact of news headlines on genetic determinism. *Science Communication*, 22(4), 379–395.

Cooley, T. B., & Lee, P. (1925). A series of cases of splenomegaly in children with anemia and peculiar bone changes. *American Journal of Diseases of Children*, 30, 447.

Correns, C. (1950/1900). G. Mendel's law concerning the behavior of progeny of varietal hybrids. *Genetics*, 35, 33–41.

Couzin-Frankel, J. (2015a). Bad luck and cancer: a science reporter's reflections on a controversial story. *Science* (available at www.sciencemag.org/news/2015/01/bad-luck-and-cancer-science-reporter-s-reflections-controversial-story).

Couzin-Frankel, J. (2015b). The bad luck of cancer. *Science*, 347(6217), 12.

Cranor, C. F. (2013). Assessing genes as causes of human disease in a multi-causal world. In Krimsky, S., & Gruber J. (Eds.) *Genetic Explanations: Sense and Nonsense*. Cambridge MA: Harvard University Press, 107–121.

Creighton, H. B., & McClintock, B. (1931). A correlation of cytological and genetical crossing-over in Zea mays. *Proceedings of the National Academy of Sciences*, 17(8), 492–497.

Crick, F. (1958). On protein synthesis. *Symposium of the Society of Experimental Biology*, 12, 138–163.

Crick, F. (1970). Central dogma of molecular biology. *Nature*, 227, 561–563.

Crick, F. (1979). Split genes and RNA splicing. *Science*, 204, 264–271.

Crick, F. H. C., Griffith, J. S., & Orgel, L. E. (1957). Codes without commas. *Proceedings of the National Academy of Sciences*, 43, 416–421.

Crick, F. H. C. (1988), *What Mad Pursuit: A Personal View of Scientific Discovery*. New York: Basic Books.

Crick, F. H. C., Barnett, L., Brenner, S., & Watts-Tobin, R. J. (1961). General nature of the genetic code for proteins. *Nature*, 192, 1227–1232.

Dar-Nimrod I. (2012). Postgenomics and genetic essentialism. *Behavioral and Brain Sciences*, 35, 362–363.

Dar-Nimrod, I., & Heine, S. J. (2011). Genetic essentialism: on the deceptive determinism of DNA. *Psychological Bulletin*, 137(5), 800–818.

Dar-Nimrod, I., Cheung, B., Ruby, M., & Heine, S. (2014). Can merely learning about obesity genes affect eating behavior? *Appetite*, 81, 269–276.

Darden, L. (1991). *Theory Change in Science: Strategies from Mendelian Genetics*. New York: Oxford University Press.

Darwin, C. (1859). *On the Origin of Species by Means of Natural Selection*. London: John Murray.

Darwin, C. (1868). *The Variation of Animals and Plants under Domestication*. London: John Murray.

Darwin, C. (1871). Pangenesis. *Nature*, 3, 502–503.

Davenport, G. C., & Davenport, C. B. (1907). Heredity of eye-color in man. *Science*, 26, 590–592.

Davies, J. A. (2014). *Life Unfolding: How the Human Body Creates Itself*. Oxford: Oxford University Press.

Davies, K. (2001). *Cracking the Genome: Inside the Race to Unlock Human DNA*. New York: Free Press.

Davies, K. (2010). *The $1,000 Genome: The Revolution in DNA Sequencing and the New Era of Personalized Medicine*. New York: Free Press.

Dawson, M. A., & Kouzarides, T. (2012). Cancer epigenetics: from mechanism to therapy. *Cell*, 150(1), 12–27.

De Vries, H. (1910/1889). *Intracellular Pangenesis*. Chicago: Open Court Publishing Co.

De Vries, H. (1950/1900). Concerning the law of segregation of hybrids. *Genetics*, 35, 30–32.

Denney, R. M., Koch, H., & Craig, I. W. (1999). Association between monoamine oxidase A activity in human male skin fibroblasts and genotype of the MAOA promoter-associated variable number tandem repeat. *Human Genetics*, 105(6), 542–551.

Depew, D. J., & Weber, B. H. (1995). *Darwinism evolving: Systems dynamics and the genealogy of natural selection*. Cambridge MA: MIT Press.

Dermitzakis, E. T. (2012). Cellular genomics for complex traits. *Nature Reviews Genetics*, 13, 215–220.

DiSilvestre, D., Koch, R., & Groffen, J. (1991). Different clinical manifestations of hyperphenylalaninemia in three siblings with identical phenylalanine hydroxylase genes. *American Journal of Human Genetics*, 48(5), 1014–1016.

Dixon, J., Jones, N. C., Sandell, L. L. et al. (2006). Tcof1/Treacle is required for neural crest cell formation and proliferation deficiencies that cause craniofacial abnormalities. *Proceedings of the National Academy of Sciences*, 103(36), 13403–13408.

Donnelly, P. (2008). Progress and challenges in genome-wide association studies in humans. *Nature*, 456(7223), 728–731.

Donovan, J., & Venville, G. (2014). Blood and bones: the influence of the mass media on Australian primary school children's understandings of genes and DNA. *Science & Education*, 23(2), 325–360.

Doolittle, W. F. (2013). Is junk DNA bunk? A critique of ENCODE. *Proceedings of the National Academy of Sciences*, 110(14), 5294–5300.

Du, J., Johnson, L. M., Jacobsen, S. E., & Patel, D. J. (2015). DNA methylation pathways and their crosstalk with histone methylation. *Nature Reviews Molecular Cell Biology*, 16(9), 519–532.

Duffy, D. L., Montgomery, G. W., Chen, W. et al. (2007). A three–single-nucleotide polymorphism haplotype in intron 1 of *OCA2* explains most human eye-color variation. *American Journal of Human Genetics*, 80(2), 241–252.

Duncan, R. G., & Reiser, B. J. (2007). Reasoning across ontologically distinct levels: students' understandings of molecular genetics. *Journal of Research in Science Teaching*, 938–959.

Dunn, L. C. (1991/1965). *A Short History of Genetics*. Ames Iowa: Iowa State University Press.

Easton, D. F., Pharoah, P. D., Antoniou, A. C. et al. (2015). Gene-panel sequencing and the prediction of breast-cancer risk. *New England Journal of Medicine*, 372(23), 2243–2257.

Ecker, U. K., Lewandowsky, S., Chang, E. P., & Pillai, R. (2014). The effects of subtle misinformation in news headlines. *Journal of Experimental Psychology: Applied*, 20(4), 323–335.

Eddy, S. R. (2013). The ENCODE project: missteps overshadowing a success. *Current Biology*, 23(7), R259–R261.

Edwards, M., Cha, D., Krithika, S., et al. (2016). Iris pigmentation as a quantitative trait: variation in populations of European, East Asian and South Asian ancestry and association with candidate gene polymorphisms. *Pigment Cell & Melanoma Research*, 29(2): 141–162.

Eiberg, H., Troelsen, J., Nielsen, M., et al. (2008). Blue eye color in humans may be caused by a perfectly associated founder mutation in a regulatory element located within the HERC2 gene inhibiting OCA2 expression. *Human Genetics*, 123, 177–187.

ENCODE Project Consortium (2007). Identification and analysis of functional elements in 1% of the human genome by the ENCODE pilot project. *Nature*, 447(7146), 799–816.

ENCODE Project Consortium (2012). An integrated encyclopedia of DNA elements in the human genome. *Nature*, 489(7414), 57–74.

Engel, E. (1972). The making of an XYY. *American Journal of Mental Deficiency*, 77(2), 123–127.

Enns, G. M., Koch, R., Brumm, V. et al. (2010). Suboptimal outcomes in patients with PKU treated early with diet alone: revisiting the evidence. *Molecular Genetics and Metabolism*, 101(2), 99–109.

Erbilgin, A., Civelek, M., Romanoski, C. E. et al. (2013). Identification of CAD candidate genes in GWAS loci and their expression in vascular cells. *Journal of Lipid Research*, 54, 1894–1905.

Esteller, M. (2008). Epigenetics in cancer. *New England Journal of Medicine*, 358(11), 1148–1159.

Esteller, M. (2011). Epigenetic changes in cancer. *F1000 Biol Rep*, 3, 9. doi: 10.3410/ B3-9

Etchegary, H., Cappelli, M., Potter, B. et al. (2010). Attitude and knowledge about genetics and genetic testing. *Public Health Genomics*, 13(2), 80–88.

Evans, J. P., Meslin, E.M., Marteau, T.M. et al. (2011). Deflating the genomic bubble. *Science*, 331, 861–862.

Falk, R. (2009). *Genetic Analysis: A History of Genetic Thinking.* Cambridge: Cambridge University Press.

Farajollahi, S., & Maas, S. (2010). Molecular diversity through RNA editing: a balancing act. *Trends in Genetics*, 26(5), 221–230.

Feinberg, A. P., & Tycko, B. (2004). The history of cancer epigenetics. *Nature Reviews Cancer*, 4, 143–153.

Feng, S., Jacobsen, S. E., & Reik, W. (2010). Epigenetic reprogramming in plant and animal development. *Science*, 330(6004), 622–627.

Fernàndez-Castillo, N., & Cormand, B. (2016). Aggressive behavior in humans: genes and pathways identified through association studies. *American Journal of Medical Genetics Part B: Neuropsychiatric Genetics*, 171(5), 676–696.

Feuk, L., Carson, A. R., & Scherer, S. W. (2006). Structural variation in the human genome. *Nature Reviews Genetics*, 7(2), 85–97.

Ficks, C. A., & Waldman, I. D. (2014). Candidate genes for aggression and antisocial behavior: a meta-analysis of association studies of the 5HTTLPR and MAOA-uVNTR. *Behavior Genetics*, 44(5), 427–444.

Fields, S., & Johnston, M. (2010). *Genetic Twists of Fate.* Cambridge MA: MIT Press.

Firenstein, S. (2012). *Ignorance: How It Drives Science.* Oxford: Oxford University Press.

Firenstein, S. (2016). *Failure: Why Science Is so Successful.* Oxford: Oxford University Press.

Følling, A. (1934). The excretion of phenylpyruvic acid in the urine, an anomaly of metabolism in connection with imbecility. *Zeitschrift fur Physiologische Chemie*, 227, 169–176.

Følling, I. (1994). The discovery of phenylketonuria. *Acta Paediatrica*, 83(s407), 4–10.

Fox, R. G. (1971). The XYY offender: a modern myth? *Journal of Criminal Law, Criminology, and Police Science*, 62(1), 59–73.

Fraga, M. F., Ballestar, E., Paz, M. F. et al. (2005). Epigenetic differences arise during the lifetime of monozygotic twins. *Proceedings of the National Academy of Sciences*, 102(30), 10604–10609.

Franklin R., & Gosling, R. G. (1953). Molecular configuration in sodium thymonucleate. *Nature*, 171, 740–741.

Frazer, K. A., Murray, S. S., Schork, N. J., & Topol, E. J. (2009). Human genetic variation and its contribution to complex traits. *Nature Reviews Genetics*, 10(4), 241–251.

Frudakis, T., Terravainen, T., & Thomas, M. (2007). Multilocus OCA2 genotypes specify human iris colors. *Human Genetics*, 122(3–4), 311–326.

Frudakis, T., Thomas, M., Gaskin, Z. et al. (2003). Sequences associated with human iris pigmentation. *Genetics*, 165(4), 2071–2083.

Fung-Leung, W. P., Schilham, M. W., Rahemtulla, A. et al. (1991). CD8 is needed for development of cytotoxic T but not helper T cells. *Cell*, 65(3), 443–449.

Galton, F. (1871a). Experiments in pangenesis, by breeding from rabbits of a pure variety, into whose circulation blood taken from other varieties had previously been largely transfused. *Proceedings of the Royal Society*, 19, 393–410.

Galton, F. (1871b). Pangenesis. *Nature*, 4, 5–6.

Galton, F. (1874). *English Men of Science: Their Nature and Nurture*. London: Macmillan & Co.

Galton, F. (1876). A theory of heredity. *Journal of the Anthropological Institute*, 5, 329–348.

Galton, F. (1886). Hereditary stature. *Nature*, 33, 295–298.

Galton, F. (1889). *Natural Inheritance*. London: Macmillan

Gannett, L. (1999). What's in a cause? The pragmatic dimensions of genetic explanations. *Biology and Philosophy*, 14, 349–373.

Gannett, L. (2013). Genomics and society: why "discovery" matters. In Kampourakis, K. (Ed.) *The Philosophy of Biology: A Companion for Educators*. Dordrecht: Springer, 653–685.

Gannett, L. (2014). "The Human Genome Project," The Stanford Encyclopedia of Philosophy (Winter 2014 Edition), Zalta, E. N. (Ed.) http://plato.stanford.edu/archives/win2014/entries/human-genome/.

Gelman, S. A. (2003). *The Essential Child: Origins of Essentialism in Everyday Thought*. Oxford: Oxford University Press.

Gelman, S. A. (2004). Psychological essentialism in children. *TRENDS in Cognitive Sciences*, 8(9), 404–409.

Gericke, N., Hagberg, M., Carvalho Santos, V., Joaquim, L. M., & El-Hani, C. (2014). Conceptual variation or incoherence? Textbook discourse on genes in six countries. *Science & Education*, 23(2), 381–416.

Gerstein, M. B., Bruce, C., Rozowsky, J. S. et al. (2007). What is a gene, post-ENCODE? History and updated definition. *Genome Research*, 17(6), 669–681.

Giardine, B., Borg, J., Higgs, D. R. et al. (2011). Systematic documentation and analysis of human genetic variation in hemoglobinopathies using the microattribution approach. *Nature Genetics*, 43(4), 295–301.

Giardine, B., Borg, J., Viennas, E. et al. (2014). Updates of the HbVar database of human hemoglobin variants and thalassemia mutations. *Nucleic Acids Research*, 42(D1), D1063–D1069.

Gibson, G. (2012). Rare and common variants: twenty arguments. *Nature Reviews Genetics*, 13(2), 135–145.

Giere, R. N. (2006) *Scientific Perspectivism*. Chicago and London: University of Chicago Press.

Gigerenzer, G. (2002). *Calculated Risks: How to Know When Numbers Deceive You*. New York: Simon and Schuster.

Gilbert, W. (1978). Why genes in pieces? *Nature*, 271, 501.

Gilbert, W. (1992). A vision of the grail. In Kevles, D. J., & Hood L. (Eds.) *The Code of Codes: Scientific and Social Issues in the Human Genome Project.* Cambridge MA: Harvard University Press, 83–97.

Gilbert, W., & Muller-Hill, B. (1966). Isolation of the *lac* repressor. *Proceedings of the National Academy of Sciences,* 56(6), 1891–1898.

Gill, P., Jeffreys, A. J., & Werrett, D. J. (1985). Forensic application of DNA "fingerprints." *Nature,* 318(6046), 577–579.

Gillham, N. W. (2001). *A Life of Sir Francis Galton: From African Explorations to the Birth of Eugenics.* Oxford: Oxford University Press.

Gladwell, M. (2008). *Outliers: The Story of Success.* New York: Little, Brown and Company.

Gliboff, S. (1998). Evolution, revolution, and reform in Vienna: Franz Unger's ideas on descent and their post-1848 reception. *Journal of the History of Biology,* 31(2), 179–209.

Gliboff, S. (2013). The many sides of Gregor Mendel. In Harman, O., & Dietrich, M. R. (Eds.) *Outsider Scientists: Routes to Innovation in Biology.* Chicago: University of Chicago Press.

Godfrey-Smith, P. (2014). *Philosophy of Biology.* Princeton: Princeton University Press.

Goding, C. R., Pei, D., & Lu, X. (2014). Cancer: pathological nuclear reprogramming?. *Nature Reviews Cancer,* 14(8), 568–573.

Gofman, J. W., Delalla, O., Glazier, F., et al. (1955). The serum lipoprotein transport system in health, metabolic disorders, atherosclerosis and coronary heart disease. *Plasma,* 2, 413–484; Reprinted as Gofman, J. W., Delalla, O., Glazier, F., et al. (2007). The serum lipoprotein transport system in health, metabolic disorders, atherosclerosis and coronary heart disease. *Journal of Clinical Lipidology,* 1(2), 104–141.

Gofman, J. W., Lindgren, F., Elliott, H., et al. (1950). The role of lipids and lipoproteins in atherosclerosis. *Science,* 111, 166–171.

Goldsmith, L., Jackson, L., O'Connor, A., & Skirton, H. (2012). Direct-to-consumer genomic testing: systematic review of the literature on user perspectives. *European Journal of Human Genetics,* 8, 811–816.

Goldstein, J. L., & Brown, M. S. (2015). A century of cholesterol and coronaries: from plaques to genes to statins. *Cell,* 161(1), 161–172.

Gott, J. M., & Emeson, R. B. (2000). Functions and mechanisms of RNA editing. *Annual Review of Genetics,* 34(1), 499–531.

Gould, S. J. (1996). *The Mismeasure of Man (Revised and Expanded Edition).* New York & London: W.W. Norton & Company.

Gould, W. A., & Heine, S. J. (2012). Implicit essentialism: genetic concepts are implicitly associated with fate concepts. *PLoS ONE,* 7(6), e38176. doi:10.1371/journal.pone.0038176.

Graur, D., Zheng, Y., & Azevedo, R. B. (2015). An evolutionary classification of genomic function. *Genome Biology and Evolution,* 7(3), 642–645.

Graur, D., Zheng, Y., Price, N. et al. (2013). On the immortality of television sets: "function" in the human genome according to the evolution-free gospel of ENCODE. *Genome Biology and Evolution,* 5(3), 578–590.

Green, R.C., & Farahany, N.A. (2014) The FDA is overcautious on consumer genomics. *Nature*, 505, 286–287.

Greenman, C., Stephens, P., Smith, R. et al. (2007). Patterns of somatic mutation in human cancer genomes. *Nature*, 446(7132), 153–158.

Griffiths, P., & Stotz, K. (2013). *Genetics and Philosophy: An Introduction.* Cambridge: Cambridge University Press.

Griffiths, A. J. F., Wessler, S. R., Lewontin, R. C. et al. (2005). *An Introduction to Genetic Analysis* (8th ed.). New York: WH Freeman & Company.

Gros, F., Hiatt, H., Gilbert, W. et al. (1961). Unstable ribonucleic acid revealed by pulse labelling of Escherichia coli. *Nature*, 190, 581–585.

Gudbjartsson, D. F., Walters, G. B., Thorleifsson, G. et al. (2008). Many sequence variants affecting diversity of adult human height. *Nature Genetics*, 40(5), 609–615.

Guthrie R., & Susi A. (1963). A simple phenylalanine method for detecting phenylketonuria in large populations of newborn infants. *Pediatrics*, 32, 338–343.

Guttmacher, A. E., Porteous, M. E., & McInerney, J. D. (2007). Educating healthcare professionals about genetics and genomics. *Nature Reviews Genetics*, 8(2), 151–157.

Haga, S., Burke W., Ginsburg G., Mills R., & Agans R. (2012). Primary care physicians' knowledge of and experience with pharmacogenomics testing. *Clinical Genetics*, 82(4), 388–394.

Haga, S. B., O'Daniel, J. M., Tindall, G. M. et al. (2012). Survey of genetic counselors and clinical geneticists' use and attitudes toward pharmacogenetic testing. *Clinical Genetics*, 82, 115–120.

Haldane, J. B. S. (1936). A provisional map of a human chromosome. *Nature*, 137, 398–400..

Haldane, J. B. S. (1938). Blood royal: a study of haemophilia in the royal families of Europe. *Modern Quarterly*, 1, 129–39.

Haldane, J. B. S. (1939). Blood royal. *The Living Age*, 26–31.

Hall, B. K. (2011). A brief history of the term and concept of epigenetics. In Hallgrímsson, B., & Hall, B. K. (Eds.) *Epigenetics: Linking Genotype and Phenotype in Development and Evolution.* Berkeley: University of California Press, 9–13.

Hall, B., Limaye, A., & Kulkarni, A. B. (2009). Overview: generation of gene knockout mice. *Current Protocols in Cell Biology*, Unit 19–12, 11–17.

Hanahan, D., & Weinberg, R. A. (2000). The hallmarks of cancer. *Cell*, 100(1), 57–70.

Hanahan, D., & Weinberg, R. A. (2011). Hallmarks of cancer: the next generation. *Cell*, 144(5), 646–674.

Hand, D. J. (2014). *The Improbability Principle: Why Coincidences, Miracles, and Rare Events Happen Day.* New York: Scientific American/Farrar, Straus and Giroux.

Harper, J. C., & SenGupta, S. B. (2012). Preimplantation genetic diagnosis: state of the art 2011. *Human Genetics*, 131(2), 175–186.

Hayden, E. C. (2016). Seeing deadly mutations in a new light. *Nature*, 538, 154–157.

Heard, E., & Martienssen, R. A. (2014). Transgenerational epigenetic inheritance: myths and mechanisms. *Cell*, 157(1), 95–109.

Hegreness, M., & Meselson, M. (2007). What did Sutton see?: Thirty years of confusion over the chromosomal basis of Mendelism. *Genetics*, 176(4), 1939–1944.

Henikoff, S., Keene, M. A., Fechtel, K., & Fristrom, J. W. (1986). Gene within a gene: nested Drosophila genes encode unrelated proteins on opposite DNA strands. *Cell*, 44, 33–42

Hershey, A. D., & Chase, M. (1952). Independent functions of viral proteins and nucleic acid in growth of bacteriophage. *Journal of General Physiology*, 36, 39–56.

Higgs, D. R., Engel, J. D., & Stamatoyannopoulos, G. (2012). Thalassaemia. *The Lancet*, 379(9813), 373–383.

Hill, R. E., & Lettice, L. A. (2013). Alterations to the remote control of *Shh* gene expression cause congenital abnormalities. *Philosphical Transactions of the Royal Society Part B*, 368, 20120357.

Hoagland, M. B., Stephenson, M. L., Scott, J. F., Hecht, L. I., & Zamecnik, P. C. (1958). A soluble ribonucleic acid intermediate in protein synthesis. *Journal of Biological Chemistry*, 231, 241–257.

Hochedlinger, K., & Plath, K. (2009). Epigenetic reprogramming and induced pluripotency. *Development*, 136(4), 509–523.

Holmes, F. L. (2000). Seymour Benzer and the definition of the gene. In Beurton, P. J., Falk, R., & Rheinberger, H.-J. (Eds.) *The Concept of the Gene in Development and Evolution*. Cambridge: Cambridge University Press, 115–155.

Holmes, S. J., & Loomis, H. M. (1909). The heredity of eye color and hair color in man. *Biological Bulletin*, 18(1), 50–65.

Holoch, D., & Moazed, D. (2015). RNA-mediated epigenetic regulation of gene 3expression. *Nature Reviews Genetics*, 16(2), 71–84.

Horowitz, N. H. (1948). The one gene-one enzyme hypothesis. *Genetics*, 33(6), 612–613.

Horton, J. D., Cohen, J. C., & Hobbs, H. H. (2009). PCSK9: a convertase that coordinates LDL catabolism. *Journal of Lipid Research*, 50, S172–S177.

Horton, W. A., Hall, J. G., & Hecht, J. T. (2007). Achondroplasia. *The Lancet*, 370, 162–172.

Hubbard, R. (2013). The mismeasure of the gene. In Krimsky, S., & Gruber, J. (Eds). *Genetic Explanations: Sense and Nonsense*. Cambridge MA: Harvard University Press, 17–25.

Hubbard, R., & Wald, E. (1997). *Exploding the Gene Myth: How Genetic Information Is Produced and Manipulated by Scientists, Physicians, Employers, Insurance Companies, Educations, and Law Enforcers*. Boston: Beacon Press.

Hughes, V. (2014). Epigenetics: the sins of the father. *Nature*, 507, 22–24.

Hurst, C. C. (1908). On the inheritance of eye-colour in man. *Proceedings of the Royal Society of London. Series B, Containing Papers of a Biological Character*, 80, 85–96.

Ingram, V. M. (1956). A specific chemical difference between globins of normal and sickle-cell anemia hemoglobins. *Nature*, 178(4537), 792–794.

Ioannidis, J. P. A. (2005). Why most published research findings are false. *PLoS Medicine*, 2(8), e124.

Ioannidis, J. P. A. (2016). Why most clinical research is not useful. *PLoS Medicine*, 13(6), e1002049.

Jablonka, E. (2013). Some problems with genetic horoscopes. In Krimsky, S., & Gruber, J. (Eds.) *Genetic Explanations: Sense and Nonsense*. Cambridge MA: Harvard University Press, 76–80.

Jablonka, E., & Lamb, M. J. (2002). The changing concept of epigenetics. *Annals of the New York Academy of Sciences*, 981(1), 82–96.

Jablonka E., & Lamb, M. J. (2014). *Evolution in Four Dimensions: Genetic, Epigenetic, Behavioral, and Symbolic Variation in the History of Life* (2nd ed). Cambridge MA: MIT Press.

Jacob, F., & Monod J. (1961). Genetic regulatory mechanisms in the synthesis of proteins. *Journal of Molecular Biology*, 3, 318–356.

Jacobs, P. A. (1982). The William Allan Memorial Award address: human population cytogenetics: the first twenty-five years. *American Journal of Human Genetics*, 34(5), 689.

Jacobs, P. A., Brunton, M., Melville, M. M., Brittain, R. P., & McClemont, W. F. (1965). Aggressive behaviour, mental sub-normality and the XYY male. *Nature*, 208, 1351–1352.

Jäger, R. J., Anvret, M., Hall, K., & Scherer, G. (1990). A human XY female with a frameshift mutation in the candidate testis-determining gene SRY. *Nature*, 348, 452–454.

Jamieson A., & Radick G. (2013). Putting Mendel in his place: how curriculum reform in genetics and counterfactual history of science can work together. In Kampourakis, K. (Ed.) *The Philosophy of Biology: A Companion for Educators*. Dordrecht: Springer, 577–595.

Jeffreys, A. J., Brookfield, J. F. Y., & Semeonoff, R. (1985). Positive identification of an immigration test-case using human DNA fingerprints. *Nature*, 317, 818–819.

Jeffreys, A. J., Wilson, V., & Thein, S. L. (1985). Individual-specific "fingerprints" of human DNA. *Nature*, 316(6023), 76–79.

Jensen, A. R. (1969). How much can we boost IQ and scholastic achievement? *Harvard Educational Review*, 39, 1–123.

Jensen, A. R. (1970). Race and the genetics of intelligence: a reply to Lewontin. *Bulletin of the Atomic Scientists*, 26(5), 17–23.

Jervis, G. A. (1937). Phenylpyruvic oligophrenia: introductory study of fifty cases of mental deficiency associated with excretion of phenylpyruvic acid. *Archives of Neurology and Psychiatry*, 38(5), 944.

Jervis, G. A. (1947). Studies on phenylpyruvic oligophrenia: the position of the metabolic error. *Journal of Biological Chemistry*, 169, 651–656.

Jervis, G. A. (1953). Phenylpyruvic oligophrenia deficiency of phenylalanine-oxidizing system. *Proceedings of the Society for Experimental Biology and Medicine*, 82(3), 514–515.

Jobling, M., & Gill, P. (2004). Encoded evidence: DNA in forensic analysis. *Nature Reviews Genetics*, 5, 739–751.

Johannsen, W. (1911). The genotype conception of heredity. *American Naturalist*, 45 (531), 129–159.

Jones, D. S. (2013). The prospects of personalized medicine. In Krimsky, S., & Gruber, J. (Eds.) *Genetic Explanations: Sense and Nonsense*. Cambridge MA: Harvard University Press, 147–170.

Judson, H. F. (1996). *The Eighth Day of Creation: The Makers of the Revolution in Biology (Commemorative Edition)*. New York: Cold Spring Harbor Laboratory Press.

Kalf, R. R., Mihaescu, R., Kundu, S. et al. (2014). Variations in predicted risks in personal genome testing for common complex diseases. *Genetics in Medicine*, 16(1), 85–91.

Kampourakis, K. (2013). Mendel and the path to genetics: portraying science as a social process. *Science & Education*, 22(2), 293–324.

Kampourakis, K. (2014). *Understanding Evolution*. Cambridge: Cambridge University Press.

Kampourakis, K. (2015). Myth 16: that Gregor Mendel was a lonely pioneer of genetics, being ahead of his time. In Numbers, R. L., & Kampourakis, K. (Eds.) *Newton's Apple and Other Myths about Science*. Cambridge MA: Harvard University Press, 129–138.

Kampourakis, K. (2016). The "general aspects" conceptualization as a pragmatic and effective means to introducing students to nature of science. *Journal of Research in Science Teaching*, 53(5), 667–682.

Kampourakis, K., Vayena, E., Mitropoulou, C. et al. (2014). Next-generation pharmacogenomics and society: key challenges ahead. *EMBO Reports*, 15(5), 472–476.

Kaplan, J.M. (2000). *The Limits and Lies of Human Genetic Research: Dangers for Social Policy*. London: Routledge.

Karlin, S., Chen, C., Gentles, A. J., & Cleary, M. (2002). Associations between human disease genes and overlapping gene groups and multiple amino acid runs. *Proceedings of the National Academy of Sciencs*, 99, 17008–17013.

Kay, L. E. (2000). *Who Wrote the Book of Life?: A History of the Genetic Code*. Stanford: Stanford University Press.

Kayaalp, E., Treacy, E., Waters, P. J. et al. (1997). Human phenylalanine hydroxylase mutations and hyperphenylalaninemia phenotypes: a metanalysis of genotype-phenotype correlations. *American Journal of Human Genetics*, 61(6), 1309–1317.

Kayser, M., & deKnijff, P. (2011). Improving human forensics through advances in genetics, genomics and molecular biology. *Nature Reviews Genetics*, 12, 179–192.

Kayser, M., Liu, F., Janssens, A. C. J. et al. (2008). Three genome-wide association studies and a linkage analysis identify *HERC2* as a human iris color gene. *American Journal of Human Genetics*, 82(2), 411–423.

Keller, E. F. (1983). *A Feeling for the Organism: The Life and Work of Barbara McClintock*. New York: Henry Holt.

Keller, E. F. (2000). *The Century of the Gene*. Cambridge MA: Harvard University Press.

Keller, E. F. (2010). *The Mirage of a Space between Nature and Nurture*. Duke University Press.

Keller, E. F. (2015). The postgenomic genome. In Stevens, H., & Richardson, S. S. (Eds.) *Postgenomics: Perspectives on Biology after the Genome*. Durham & London: Duke University Press, 9–31.

Keren, H., Lev-Maor, G., & Ast, G. (2010). Alternative splicing and evolution: diversification, exon definition and function. *Nature Reviews Genetics*, 11(5), 345–355.

Kevles, D. J. (1995). *In the Name of Eugenics: Genetics and the Uses of Human Heredity*. Cambridge MA: Harvard University Press.

Keys, A., Kimura, N., Kusukawa, A. et al. (1958). Lessons from serum cholesterol studies in Japan, Hawaii and Los Angeles. *Annals of Internal Medicine*, 48(1), 83–94.

Kioussis, D., Vanin, E., DeLange, T., Flavell, R. A., & Grosveld, F. G. (1983). β-globin gene inactivation by DNA translocation in γ β-thalassaemia. *Nature*, 306, 662–666.

Kirby, D. A. (2000). The new eugenics in cinema: genetic determinism and gene therapy in GATTACA. *Science Fiction Studies*, 27(2), 193–215.

Kirby, D. A. (2004). Extrapolating race in GATTACA: genetic passing, identity, and the science of race. *Literature and Medicine*, 23(1), 184–200.

Kitcher, P. (1982). Genes. *British Journal for the Philosophy of Science*, 33, 337–359.

Kitcher, P. (1997). *The Lives to Come: The Genetic Revolution and Human Possibilities*. New York: Touchstone.

Kitcher, P. (2003). Battling the undead: how (and how not) to resist genetic determinism. In Kitcher, P. *In Mendel's Mirror: Philosophical Reflections on Biology*. Oxford: Oxford University Press, 283–300.

Klitzman, R. L. (2012). *Am I My Genes?: Confronting Fate and Family Secrets in the Age of Genetic Testing*. Oxford and New York: Oxford University Press.

Knoll, J. H. M., Nicholls, R. D., Magenis, R. E. et al. (1989). Angelman and Prader-Willi syndromes share a common chromosome 15 deletion but differ in parental origin of the deletion. *American Journal of Medical Genetics*, 32(2), 285–290.

Kohler, R. E. (1994). *Lords of the Fly: Drosophila Genetics and the Experimental Life*. Chicago: University of Chicago Press.

Koller, B. H., Marrack, P., Kappler, J. W., & Smithies, O. (1990). Normal development of mice deficient in β2M, MHC class I proteins, and CD8+ T cells. *Science*, 248, 1227–1230.

Kong, A., Frigge, M. L., Masson, G. et al. (2012). Rate of de novo mutations and the importance of father's age to disease risk. *Nature*, 488(7412), 471–475.

Kornberg, R. D. (1974). Chromatin structure: a repeating unit of histones and DNA. *Science*, 184, 868–871.

Kornberg, R. D., & Thomas, J. O. (1974). Chromatin structure; oligomers of the histones. *Science*, 184, 865–868.

Kouzarides, T. (2007). Chromatin modifications and their function. *Cell*, 128(4), 693–705.

Krimsky S., & Gruber J. (Eds.) (2013). *Genetic Explanations: Sense and Nonsense*. Cambridge MA: Harvard University Press.

Krimsky, S., & Simoncelli, T. (2011). *Genetic Justice: DNA Data Banks, Criminal Investigations, and Civil Liberties*. New York: Columbia University Press.

Kruglyak, L. (2008). The road to genome-wide association studies. *Nature Reviews Genetics*, 9(4), 314–318.

Kurian, A. W., Hare, E. E., Mills, M. A. et al. (2014). Clinical evaluation of a multiple-gene sequencing panel for hereditary cancer risk assessment. *Journal of Clinical Oncology*, 32(19), 2001–2009.

Lachance, C. R., Erby, L. A., Ford, B. M., Allen, V. C., & Kaphingst, K. A. (2010). Informational content, literacy demands, and usability of websites offering health-related genetic tests directly to consumers. *Genetics in Medicine*, 12(5), 304–312.

Lander, E. S. (2011). Initial impact of the sequencing of the human genome. *Nature*, 470(7333), 187–197.

Lander, E. S. (2015). Cutting the Gordian helix–regulating genomic testing in the era of precision medicine. *New England Journal of Medicine*, 372(13), 1185–1186.

Lander, E. S., Linton, L. M., Birren, B. et al. (2001). Initial sequencing and analysis of the human genome. *Nature*, 409(6822), 860–921.

Lannoy, N., & Hermans, C. (2010). The "royal disease": haemophilia A or B? A haematological mystery is finally solved. *Haemophilia*, 16(6), 843–847.

Lee, J. T., & Bartolomei, M. S. (2013). X-inactivation, imprinting, and long noncoding RNAs in health and disease. *Cell*, 152(6), 1308–1323.

Leighton J. W., Valverde K., & Bernhardt B. A. (2012). The general public's understanding and perception of direct-to-consumer genetic test results. *Public Health Genomics*. 15(1), 11–21.

Lek, M., Karczewski, K., Minikel, E., Samocha, K., Banks, E., Fennell, T., et al. (2016). Analysis of protein-coding genetic variation in 60,706 humans. *Nature*, 536, 285–291.

Lettice, L. A., Horikoshi, T., Heaney, S. J. H. et al. (2002). Disruption of a long-range cis-acting regulator for *Shh* causes preaxial polydactyly. *Proceedings of the National Academy of Sciences*, 99, 7548–7553.

Lettre, G., Jackson, A. U., Gieger, C. et al. (2008). Identification of ten loci associated with height highlights new biological pathways in human growth. *Nature Genetics*, 40(5), 584–591.

Levins, R., & Lewontin, R. C. (1985). *The Dialectical Biologist*. Cambridge MA: Harvard University Press.

Levy, S., Sutton, G., Ng, P. C. et al. (2007). The diploid genome sequence of an individual human. *PLoS Biology* 5(10), e254.

Lewin, B. (1980). Alternatives for splicing: recognizing the ends of introns. *Cell*, 22, 324–326.

Lewis, J. D., Meehan, R. R., Henzel, W. J. et al. (1992). Purification, sequence, and cellular localization of a novel chromosomal protein that binds to methylated DNA. *Cell*, 69(6), 905–914.

Lewis, J., Leach, J., & Wood-Robinson, C. (2000). All in the genes? – young people's understanding of the nature of genes. *Journal of Biological Education*, 34(2), 74–79.

Lewontin, R. C. (1970a). Further remarks on race and the genetics of intelligence. *Bulletin of the Atomic Scientists*, 26(5), 23–25.

Lewontin, R. C. (1970b). Race and intelligence. *Bulletin of Atomic Scientists*, 26, 2–8.

Lewontin, R. C. (1974). The analysis of variance and the analysis of causes. *American Journal of Human Genetics*, 26, 400–411.

Lewontin, R. C. (1993). *The Doctrine of DNA: Biology as Ideology.* London: Penguin.

Lewontin, R. C. (2000). *The Triple Helix: Gene, Organism, and Environment.* Cambridge MA: Harvard University Press.

Ley, B. L., Jankowski N., & Brewer P. R. (2012) Investigating *CSI*: portrayals of DNA testing on a forensic crime show and their potential effects. *Public Understanding of Science*, 21 (1), 51–67.

Libby, P., Ridker, P. M., & Hansson, G. K. (2011). Progress and challenges in translating the biology of atherosclerosis. *Nature*, 473, 317–325.

Lichtenstein, P., Holm, N. V., Verkasalo, P. K. et al. (2000). Environmental and heritable factors in the causation of cancer – analyses of cohorts of twins from Sweden, Denmark, and Finland. *New England Journal of Medicine*, 343(2), 78–85.

Lindee, S. (2005) *Moments of Truth in Genetic Medicine.* Baltimore: Johns Hopkins University Press.

Link, K. P. (1959). The discovery of dicumarol and its sequels. *Circulation*, 19(1), 97–107.

Lipkin, S. M., & Luoma, J. (2016). *The Age of Genomes: Tales from the Front Lines of Genetic Medicine.* Boston: Beacon Press.

López Beltrán, C. (2004). In the cradle of heredity: French physicians and l'hérédité naturelle in the early nineteenth century. *Journal of the History of Biology*, 37, 39–72.

López-Beltrán, C. (2007). The medical origins of heredity. In Müller-Wille, S. & Rheinberger, H. (Eds.) *Heredity Produced: At the Crossroad of Biology, Politics and Culture, 1500 to 1870.* Cambridge MA: MIT Press.

Loukopoulos, D. (2014). Milestones in the history of thalassemia and sickle cell disease. *Thalassemia Reports*, 4(3), 29–32.

Lupski, J. R. (2013). Genome mosaicism: one human, multiple genomes. *Science*, 341(6144), 358–359.

Lynch, K. W. (2004). Consequences of regulated pre-mRNA splicing in the immune system. *Nature Reviews Immunology*, 4(12), 931–940.

Lynch, M., Cole, S. A., McNally, R., & Jordan, K. (2008). *Truth Machine: The Contentious History of DNA Fingerprinting.* Chicago: University of Chicago Press.

MacArthur, D. G., Balasubramanian, S., Frankish, A. et al. (2012). A systematic survey of loss-of-function variants in human protein-coding genes. *Science*, 335, 823–828.

MacArthur, D. G., Manolio, T. A., Dimmock, D. P. et al. (2014). Guidelines for investigating causality of sequence variants in human disease. *Nature*, 508(7497), 469–476.

Maienschein, J. (2012). "Epigenesis and Preformationism." *The Stanford Encyclopedia of Philosophy* (Spring 2012 Edition), Edward N. Zalta (Ed.), http://plato.stanford.edu/archives/spr2012/entries/epigenesis/.

Maienschein, J. (2014). *Embryos under the Microscope: The Diverging Meanings of Life.* Cambridge MA: Harvard University Press.

Mak, T. W., Penninger, J. M., & Ohashi, P. S. (2001). Knockout mice: a paradigm shift in modern immunology. *Nature Reviews Immunology*, 1(1), 11–19.

Makalowska, I., Lin, C. F., & Makalowski, W. (2005). Overlapping genes in vertebrate genomes. *Computational Biology and Chemistry*, 29(1), 1–12.

Margarit, E., Coll, M. D., Oliva, R. et al. (2000). SRY gene transferred to the long arm of the X chromosome in a Y-positive XX true hermaphrodite. *American Journal of Medical Genetics*, 90(1), 25–28.

Marks, J. (2008). The construction of Mendel's laws. *Evolutionary Anthropology*, 17, 250–253.

Mathews, D. J. H., Kalfoglou, A., & Hudson, K. (2005). Geneticists' views on science policy formation and public outreach. *American Journal of Medical Genetics*, 137A, 161–169.

Mattick, J. S., & Dinger, M. E. (2013). The extent of functionality in the human genome. *HUGO Journal*, 7(1), 2.

McCarthy, M. I., Abecasis, G. R., Cardon, L. R. et al. (2008). Genome-wide association studies for complex traits: consensus, uncertainty and challenges. *Nature Reviews Genetics*, 9(5), 356–369.

McClintock, B. (1953). Induction of instability at selected loci in maize. *Genetics*, 38(6), 579.

McClintock, B. (1956). Controlling elements and the gene. *Cold Spring Harbor Symposia on Quantitative Biology*, 21, 197–216.

McClintock, B. (1961). Some parallels between gene control systems in maize and in bacteria. *American Naturalist*, 95 (884), 265–277.

Meli, C., Garozzo, R., Mollica, F., Romano, V., & Cali, F. (1998). A20-Different clinical manifestations in siblings with identical phenylalanine hydroxylase gene mutations. *Journal of Inherited Metabolic Disease*, 21(2), 10.

Mendel, G. (1866). *Versuche über Pflanzen-Hybriden*, translation by Kersten Hall and Staffan Müller-Wille, available at http://centimedia.org/bshs-translations/mendel.

Meselson, M., & Stahl, F. W. (1958). The replication of DNA in Escherichia coli. *Proceedings of the National Academy of Sciences*, 44, 671–682.

Miko, I. (2008). Gregor Mendel and the principles of inheritance. *Nature Education*, 1(1), 134.

Miller, K. F., Smith, C. M., Zhu, J., & Zhang, H. (1995). Preschool origins of cross-national differences in mathematical competence: the role of number-naming systems. *Psychological Science*, 6(1), 56–60.

Miller, K. R., & Levine, J. S. (2010). *Miller & Levine Biology*. USA: Pearson Education.

Minikel, E. V., Vallabh, S. M., Lek, M., Estrada, K., Samocha, K. E., Sathirapongsasuti, J. F., et al. (2016). Quantifying prion disease penetrance using large population control cohorts. *Science Translational Medicine*, 8(322), 322–329.

Mills Shaw, K. R., Van Horne, K., Zhang, H., & Boughman, J. (2008). Essay contest reveals misconceptions of high school students in genetics content. *Genetics*, 178(3), 1157–1168.

Moore, D. S. (2002). *The Dependent Gene: The Fallacy of "Nature vs. Nurture."* New York: Times Books/Henry Holt & Co.

Moore, D. S. (2008). Espousing interactions and fielding reactions: addressing lay-people's beliefs about genetic determinism. *Philosophical Psychology*, 21(3), 331–348.

Moore, D. S. (2013a). Big B, little b: myth #1 is that Mendelian genes actually exist. In Krimsky, S., & Gruber, J. (Eds.) *Genetic Explanations: Sense and Nonsense*. Cambridge MA: Harvard University Press, 43–50.

Moore, D. S. (2013b). Current thinking about nature and nurture. In Kampourakis, K. (Ed.) *The Philosophy of Biology: A Companion for Educators*. Dordrecht: Springer, 629–652.

Moore, D. S. (2015). *The Developing Genome: An Introduction to Behavioral Epigenetics*. Oxford: Oxford University Press.

Morange, M. (1998). *A History of Molecular Biology*. Cambridge MA: Harvard University Press.

Morange, M. (2002). *The Misunderstood Gene*. Cambridge MA: Harvard University Press.

Morell, V. (1993). Evidence found for a possible aggression gene. *Science*, 260(5115), 1722–1723.

Morgan, T. H. (1913). Factors and unit characters in Mendelian heredity. *American Naturalist*, 47, 5–16.

Morgan, T. H. (1917). The theory of the gene. *American Naturalist*, 51, 513–544.

Morgan, T. H. (1926). *The Theory of the Gene*. New Haven: Yale University Press.

Morgan, T. H., Sturtevant, A. H., Muller, H. J., & Bridges, C. B. (1915). *The Mechanism of Mendelian Heredity*. New York: Henry Holt and Company.

Morin-Chassé, A. (2014). Public (mis)understanding of news about behavioral genetics research: a survey experiment. *BioScience*, 64(12), 1170–1177.

Morris, C., Shen, A., Pierce, K., & Beckwith, J. (2007). Deconstructing violence. *GeneWatch*, 20(2), 3–9.

Morris, K. V., & Mattick, J. S. (2014). The rise of regulatory RNA. *Nature Reviews Genetics*, 15(6), 423.

Moss, L. (2003). *What Genes Can't Do*. Cambridge MA: MIT Press.

Muela, F. J., & Abril, A. M. (2014) Genetics and cinema: personal misconceptions that constitute obstacles to learning. *International Journal of Science Education, Part B: Communication and Public Engagement*, 4(3), 260–280.

Müller, C. (1938). Xanthomata, hypercholesterolemia, angina pectoris. *Acta Medica Scandinavica*, 95(S89), 75–84.

Muller, H. J. (1927). Artificial transmutation of the gene. *Science*, 46, 84–87.

Müller-Wille, S., & Rheinberger, H-J. (2012). *A Cultural History of Heredity*. Chicago: University of Chicago Press.

Mukherjee, S. (2016). *The Gene: An Intimate History*. New York: Scribner.

Murray, A. B. V., Carson, M. J., Morris, C. A., & Beckwith, J. (2010). Illusions of scientific legitimacy: misrepresented science in the direct-to-consumer genetic testing marketplace. *Trends in Genetics*, 26, 459–461.

Nägeli von, C. (1898/1884). *A Mechanico-Physiological Theory of Organic Evolution*. Chicago: Open Court Publishing Co.

Neel, J. V. (1949). The inheritance of sickle cell anemia. *Science*, 110, 64–66.

Nelkin, D. (1995). *Selling Science: How the Press Covers Science and Technology (Revised Edition)*. New York: W.H. Freeman & Company.

Nelkin, D., & Lindee, S. M. (2004). *The DNA Mystique: The Gene as a Cultural Icon*. Ann Arbor: University of Michigan Press.

Nersessian, N. J. (2008). *Creating Scientific Concepts*. Cambridge MA: MIT Press.

Neumann-Held, E. M. (2006). Genes – causes – codes: deciphering DNA's ontological privilege. In Neumann-Held, E. M., & Rehmann-Sutter, C. (Eds.) *Genes in Development: Re-Reading the Molecular Paradigm*. Durham and London: Duke University Press, 238–271.

Neumann-Held, E. M., & Rehmann-Sutter, C. (Eds.) (2006). *Genes in Development: Re-Reading the Molecular Paradigm*. Durham and London: Duke University Press.

Ng, P. C., Murray, S. S., Levy, S., & Venter, J. C. (2009). An agenda for personalized medicine. *Nature*, 461(7265), 724–726.

Ng, S. B., Bigham, A. W., Buckingham, K. J. et al. (2010). Exome sequencing identifies MLL2 mutations as a cause of Kabuki syndrome. *Nature Genetics*, 42(9), 790–793.

Nicholls, R. D., Knoll, J. H., Butler, M. G., Karam, S., & Lalande, M. (1989). Genetic imprinting suggested by maternal heterodisomy in non-deletion Prader-Willi syndrome. *Nature*, 342(6247), 281–285.

Nilsen, T. W., & Graveley, B. R. (2010). Expansion of the eukaryotic proteome by alternative splicing. *Nature*, 463 (7280), 457–463.

Nirenberg, M., Leder, P., Bernfield, M. et al. (1965). RNA codewords and protein synthesis, VII: on the general nature of the RNA code. *Proceedings of the National Academy of Sciences*, 53, 1161–1168.

Nirenberg, M. W., & Matthaei, H. J. (1961). The dependence of cell-free protein synthesis in *E. coli* upon naturally occurring or synthetic polyribonucleotides. *Proceedings of the National Academy of Sciences*, 47 (10), 1588–1602.

Nishikura, K. (2016). A-to-I editing of coding and non-coding RNAs by ADARs. *Nature Reviews Molecular Cell Biology*, 17, 83–96.

Noble, D. (2006). *The Music of Life: Biology beyond Genes*. Oxford: Oxford University Press.

Nowicki, S. (2012). *Holt McDougal Biology*. Holt McDougal.

Nuinoon, M., Makarasara, W., Mushiroda, T. et al. (2010). A genome-wide association identified the common genetic variants influence disease severity in β0-thalassemia/hemoglobin E. *Human Genetics*, 127(3), 303–314.

Núñez-Farfán, J., & Schlichting, C. D. (2001). Evolution in changing environments: the "synthetic" work of Clausen, Keck, and Hiesey. *Quarterly Review of Biology*, 76(4), 433–457.

Ogilvie, M. B., & Choquette, C. J. (1981). Nettie Maria Stevens (1861–1912): her life and contributions to cytogenetics. *Proceedings of the American Philosophical Society*, 125(4), 292–311.

Ohno, S. (1972). So much "junk" DNA in our genome. *Brookhaven Symposium on Biology*, 23, 366–370.

Olby, R. (1970). Francis Crick, DNA, and the central dogma. *Daedalus*, 99(4), 938–987.

Olby, R. (1994/1974). *The Path to the Double Helix.* New York: Dover.

Olby, R. (2003). Quiet debut for the double helix. *Nature,* 421, 402–405.

Olby, R. (2009). *Francis Crick, Hunter of Life's Secrets.* New York: Cold Spring Harbor Laboratory Press.

Olby, R. C. (1966). *Origins of Mendelism.* New York: Schocken Books.

Olby, R. C. (1979). Mendel no Mendelian? *History of Science,* 17, 53–72.

Olby, R. C. (1985). *Origins of Mendelism* (2nd ed.). Chicago: University of Chicago Press.

Olby, R. C. (2000). Horticulture: the font for the baptism of Genetics. *Nature Reviews Genetics,* 1, 65–70.

Old, J. M. (2003). Screening and genetic diagnosis of haemoglobin disorders. *Blood Reviews,* 17(1), 43–53.

Olesko, K. M. (2015). Myth 25: that science has been largely a solitary enterprise. In Numbers, R. L., & Kampourakis, K. (Eds.) *Newton's Apple and Other Myths about Science.* Cambridge MA: Harvard University Press, 202–209.

Ollikainen, M., Smith, K. R., Joo, E. J. H. et al. (2010). DNA methylation analysis of multiple tissues from newborn twins reveals both genetic and intrauterine components to variation in the human neonatal epigenome. *Human Molecular Genetics,* 19(21), 4176–4188.

Orel, V. (1984) *Mendel.* New York: Oxford University Press.

Orel, V., & Wood, R. J. (2000). Essence and origin of Mendel's discovery. *Comptes Rendus de l'Académie des Sciences, Series III Sciences de la Vie,* 323, 1037–1041.

Pace, L. E., & Keating, N. L. (2014). A systematic assessment of benefits and risks to guide breast cancer screening decisions. *Journal of the American Medical Association,* 311(13), 1327–1335.

Palade, G. E. (1955). A small particulate component of the cytoplasm. *Journal of Biophysical and Biochemical Cytology,* 1, 59–68.

Palazzo, A. F., & Gregory, T. R. (2014) The case for junk DNA. *PLoS Genetics,* 10(5), e1004351.

Pan, Q., Shai, O., Lee, L. J., Frey, B. J., Blencowe, B. J. (2008). Deep surveying of alternative splicing complexity in the human transcriptome by high-throughput sequencing. *Nature Genetics,* 40, 1413–1415.

Parrington, J. (2015). *The Deeper Genome: Why There Is More to the Human Genome than Meets the Eye.* Oxford: Oxford University Press.

Paul, D. B. (1995). *Controlling Human Heredity: 1865 to the Present.* New York: Humanity Books.

Paul, D. B. (1998). *The Politics of Heredity: Essays on Eugenics, Biomedicine, and the Nature-Nurture Debate.* New York: SUNY Press.

Paul, D. B. (2014). What was wrong with eugenics? Conflicting narratives and disputed interpretations. *Science & Education,* 23(2), 259–271.

Paul, D. B., & Brosco, J. P. (2013). *The PKU Paradox: A Short History of a Genetic Disease.* Baltimore: Johns Hopkins Universtiy Press.

Pauling, L., Itano, H. A., Singer, S. J., & Wells, I. C. (1949). Sickle cell anemia, a molecular disease. *Science,* 110 (2865), 543–548.

Pennisi, E. (2001). The human genome. *Science*, 291(5507), 1177–1180.

Penrose, L. S. (1935). Inheritance of phenylpyruvic amentia (phenylketonuria). *Lancet*, 2, 192–194.

Penrose, L. S., & Quastel, J. H. (1937). Metabolic studies in phenylketonuria. *Biochemical Journal*, 31, 266–271.

Peters, J. (2014). The role of genomic imprinting in biology and disease: an expanding view. *Nature Reviews Genetics*, 15(8), 517–530.

Pigliucci, M. (2001). *Phenotypic Plasticity: Beyond Nature and Nurture*. Baltimore: Johns Hopkins University Press.

Pigliucci, M. (2005). Evolution of phenotypic plasticity: where are we going now?. *Trends in Ecology & Evolution*, 20(9), 481–486.

Pontecorvo, G. (1952). The genetic formulation of gene structure and action. *Advances in Enzymology*, 13, 121–149.

Prader, A., Labhart, A., & Willi, H. (1956). A syndrome with adiposity, stunted growth, cryptocordia and oligophrenia after myotonia entitled in newborn. *Schweiz Med Wochenschr*, 86, 1260–1261.

Preußer, C., & Bindereif, A. (2013). Exo-endo trans splicing: a new way to link. *Cell Research*, 23(9), 1071.

Ptashne, M. (1967). Isolation of the λ phage repressor. *Proceedings of the National Academy of Sciences*, 57(2), 306–313.

Qiu, J. (2006). Epigenetics: unfinished symphony. *Nature*, 441(7090), 143–145.

Reece, J. B., Urry, L. A., Cain, M. L. et al. (2012). *Campbell Biology* (9th ed.). New York: Pearson Education.

Reid, J. B., & Ross, J. J. (2011). Mendel's genes: toward a full molecular characterization. *Genetics*, 189(1), 3–10.

Renwick C. (2011). From political economy to sociology: Francis Galton and the social-scientific origins of eugenics. *British Journal for the History of Science*, 44(162 Pt 3), 343–369.

Resch, B. (Ed.). (2011). *BSCS Biology: A Human Approach* (4th ed.). Dubuque IA: Kendall Hunt.

Retzbach, J., Retzbach, A., Maier, M., Otto, L., & Rahnke, M. (2013). Effects of repeated exposure to science TV shows on beliefs about scientific evidence and interest in science. *Journal of Media Psychology*, 25(1), 3–13.

Reydon, T. R., Kampourakis, K., & Patrinos, G. P. (2012). Genetics, genomics and society: the responsibilities of scientists for science communication and education. *Personalized Medicine*, 9(6), 633–643.

Rheinberger, H. J., & Müller-Wille, S. (in press) *The Gene: From Genetics to Postgenomics*. Chicago: University of Chicago Press.

Rheinberger, H. J., Müller-Wille, S., & Meunier, R. (2015). "Gene." The Stanford Encyclopedia of Philosophy (Spring 2015 Edition), Edward N. Zalta (Ed.), http://plato.stanford.edu/archives/spr2015/entries/gene/.

Richards, R. A. (2010). *The Species Problem: A Philosophical Analysis*. Cambridge: Cambridge University Press.

Roberts, H. F. (1929) *Plant Hybridisation Before Mendel*. Princeton: Princeton University Press.

Roll-Hansen, N. (2014). Commentary: Wilhelm Johannsen and the problem of heredity at the turn of the 19th century. *International Journal of Epidemiology*, 43(4), 1007–1013.

Rollin, B. E. (2006). *Science and Ethics*. Cambridge: Cambridge University Press.

Rose, H., & Rose, S. (2012). *Genes, Cells and Brains: The Promethean Promises of the New Biology*. London & New York: Verso.

Rothbart, S. B., & Strahl, B. D. (2014). Interpreting the language of histone and DNA modifications. *Biochimica et Biophysica Acta (BBA)-Gene Regulatory Mechanisms*, 1839(8), 627–643.

Sabol, S. Z., Hu, S., & Hamer, D. (1998). A functional polymorphism in the monoamine oxidase A gene promoter. *Human Genetics*, 103(3), 273–279.

Sadava, D., Hillis, D. M., Heller, H. C., & M. Berenbaum (2011). *Life: The Science of Biology*. Gordonsville: WH Freeman Publishers.

Sakai. D., & Trainor, P. A. (2009) Treacher Collins syndrome: unmasking the role of Tcof1/treacle. *The International Journal of Biochemistry & Cell Biology*, 41(6), 1229–1232.

Sapp, J. (2003). *Genesis: The Evolution of Biology*. Oxford: Oxford University Press.

Sarkar, S. (2006). From genes as determinants to DNA as resource: historical notes on development and genetics. In Neumann-Held, E. M., & Rehmann-Sutter, C. (Eds.) *Genes in Development: Re-Reading the Molecular Paradigm*. Durham and London: Duke University Press, 77–95.

Sato, Y., Morita, R., Nishimura, M., Yamaguchi, H., & Kusaba, M. (2007). Mendel's green cotyledon gene encodes a positive regulator of the chlorophyll-degrading pathway. *Proceedings of the National Academy of Sciences*, 104(35), 14169–14174.

Schmidt, J. L., Castellanos-Brown, K., Childress, S. et al. (2012). The impact of false-positive newborn screening results on families: a qualitative study. *Genetics in Medicine*, 14(1), 76–80.

Schmucker, D., Clemens, J. C., Shu, H., et al. (2000). Drosophila *Dscam* is an axon guidance receptor exhibiting extraordinary molecular diversity. *Cell*, 101, 671–684.

Schoenfeld, J. D., & Ioannidis, J. P. (2013). Is everything we eat associated with cancer? A systematic cookbook review. *American Journal of Clinical Nutrition*, 97(1), 127–134.

Schulte, J., Rothaus, C. S., Adler, J. N., & Phimister, E. G. (2014). Screening an asymptomatic person for genetic risk – polling results. *New England Journal of Medicine*, 371(20), 2442–2445.

Schwartz, S. (2000). The differential concept of the gene: past and present. In Beurton, P. J., Falk, R. & Rheinberger, H. J. (Eds.) *The Concept of the Gene in Development and Evolution*. Cambridge: Cambridge University Press, 26–39.

Schweitzer, N. J., & Saks, M. J. (2007). The *CSI* Effect: popular fiction about forensic science affects the public's expectations about real forensic science. *Jurimetrics*, 47, 357–364.

Scotti, M., & Swanson, M. S. (2016). RNA mis-splicing in disease. *Nature Reviews Genetics*, 17(1): 19–32.

Scriver, C. R. (2007). The PAH gene, phenylketonuria, and a paradigm shift. *Human Mutation*, 28(9), 831–845.

Scriver, C. R., & Waters, P. J. (1999). Monogenic traits are not simple: lessons from phenylketonuria. *Trends in Genetics*, 15(7), 267–272.

Seisenberger, S., Peat, J. R., Hore, T. A. et al. (2013). Reprogramming DNA methylation in the mammalian life cycle: building and breaking epigenetic barriers. *Philosophical Transactions of the Royal Society B: Biological Sciences* 368(1609), 20110330.

Sekido, R., & Lovell-Badge, R. (2009). Sex determination and SRY: down to a wink and a nudge? *Trends in Genetics*, 25(1), 19–29.

Sermon, K., Van Steirteghem, A., & Liebaers, I. (2004). Preimplantation genetic diagnosis. *The Lancet*, 363(9421), 1633–1641.

Shaheen, R., Faqeih, E., Ansari, S. et al. (2014). Genomic analysis of primordial dwarfism reveals novel disease genes. *Genome Research*, 24(2), 291–299.

Sharp, P. A. (2005). The discovery of split genes and RNA splicing. *Trends in Biochemical Sciences*, 30(6), 279–280.

Shull, G. H. (1935). The word "allele." *Science*, 82 (2115), 37–38.

Slack, J. (2014). *Genes: A Very Short Introduction*. Oxford: Oxford University Press.

Slatkin, M. (2008). Linkage disequilibrium – understanding the evolutionary past and mapping the medical future. *Nature Reviews Genetics*, 9(6), 477–485.

Smith, I., & Lloyd, J. (1974). Atypical phenylketonuria accompanied by a severe progressive neurological illness unresponsive to dietary treatment. *Archives of Disease in Childhood*, 49(3), 245.

Smith, I., Clayton, B. E., & Wolff, O. H. (1975). New variant of phenylketonuria with progressive neurological illness unresponsive to phenylalanine restriction. *The Lancet*, 305(7916), 1108–1111.

Snyder, M. (2016). *Genomics and Personalized Medicine: What Everyone Needs to Know*. Oxford: Oxford University Press.

Snyder, M., & Gerstein M. (2003). Defining genes in the genomics era. *Science*, 300, 258–260.

Sober, E., & Lewontin, R. C. (1982). Artifact, cause and genic selection. *Philosophy of Science*, 49(2),157–180.

Söll, D., Ohtsuka, E., Jones, D. S., et al. (1965). Studies on polynucleotides, XLIX. Stimulation of the binding of aminoacyl-sRNA's to ribosomes by ribotrinucleotides and a survey of codon assignments for 20 amino acids. *Proceedings of the National Academy of Sciences*, 54, 1378–1385.

Solovieff, N., Cotsapas, C., Lee, P. H., Purcell, S. M., & Smoller, J. W. (2013). Pleiotropy in complex traits: challenges and strategies. *Nature Reviews Genetics*, 14(7), 483–495.

Sommer, B., Köhler, M., Sprengel, R., & Seeburg, P. H. (1991). RNA editing in brain controls a determinant of ion flow in glutamate-gated channels. *Cell*, 67(1), 11–19.

Sonnenschein, C., & Soto, A. M. (2013). Cancer genes: the vestigial remains of a fallen theory. In Krimsky, S., & Gruber J. (Eds.) *Genetic Explanations: Sense and Nonsense*. Cambridge MA: Harvard University Press, 81–93.

Spencer, H. (1864). *Principles of Biology*. London and Edinburgh: Williams and Norgate.

Stamatoyannopoulos, J. A. (2012). What does our genome encode? *Genome Research*, 22, 1602–1611.

Stanek, E. J., Sanders, C. L., Johansen Taber, K. A. et al. (2012). Adoption of pharmacogenomic testing by US physicians: results of a nationwide survey. *Clinical Pharmacology & Therapy*, 91(3), 450–458.

Stevens, H., & Richardson, S. S. (Eds.) (2015a). Beyond the genome. In Stevens, H., & Richardson, S. S. (Eds.) *Postgenomics: Perspectives on Biology after the Genome*. Durham and London: Duke University Press, 1–8.

Stevens, H., & Richardson, S. S. (2015b). *Postgenomics: Perspectives on Biology after the Genome*. Durham & London: Duke University Press.

Stotz, K., Griffiths, P. E., & Knight, R. (2004). How biologists conceptualize genes: an empirical study. *Studies in History and Philosophy of Science Part C: Studies in History and Philosophy of Biological and Biomedical Sciences* 35(4), 647–673.

Strasser, B. J. (1999). Sickle cell anemia, a molecular disease. *Science*, 286 (5444), 1488–1490

Strasser, B. J. (2003). Who cares about the double helix? *Nature*, 422, 803–804.

Strasser, B. J. (2006). A world in one dimension: Linus Pauling, Francis Crick and the central dogma of molecular biology. *History and Philosophy of the Life Sciences*, 28, 491–512.

Strasser, B. J. (2015). Myth 22: that Linus Pauling's discovery of the molecular basis of sickle-cell anemia revolutionized medical practice. In Numbers, R. L. & Kampourakis, K. (Eds.) *Newton's Apple and Other Myths about Science*. Cambridge MA: Harvard University Press, 178–185.

Stratton, M. R., Campbell, P. J., & Futreal, P. A. (2009). The cancer genome. *Nature*, 458(7239), 719–724.

Sturm, R. A., & Frudakis, T. N. (2004). Eye colour: portals into pigmentation genes and ancestry. *TRENDS in Genetics*, 20(8), 327–332.

Sturm, R. A., & Larsson, M. (2009). Genetics of human iris colour and patterns. *Pigment Cell & Melanoma Research*, 22(5), 544–562.

Sturm, R. A., Duffy, D. L., Zhao, Z. Z. et al. (2008). A single SNP in an evolutionary conserved region within intron 86 of the *HERC2* gene determines human blue-brown eye color. *American Journal of Human Genetics*, 82, 424–431.

Sturtevant, A. H. (1913). The linear arrangement of six sex-linked factors in Drosophila, as shown by their mode of association. *Journal of Experimental Zoology*, 14, 43–59.

Sturtevant, A. H. (2001/1965). *A History of Genetics*. New York: Electronic Scholarly Publishing project & Cold Spring Harbor Laboratory Press (available at www.esp.org).

Sulem, P., Gudbjartsson, D. F., Stacey, S. N. et al. (2007). Genetic determinants of hair, eye and skin pigmentation in Europeans. *Nature Genetics* 39(12), 1443–1452.

Sulem, P., Helgason, H., Oddson, A. et al. (2015). Identification of a large set of rare complete human knockouts. *Nature Genetics*, 47(5), 448–452.

Sullivan, L. G. (1995). Myth, metaphor and hypothesis: how anthropomorphism defeats science. *Philosophical Transactions: Biological Sciences*, 349(1328), 215–218.

Sun, J. X., Helgason, A., Masson, G. et al. (2012). A direct characterization of human mutation based on microsatellites. *Nature Genetics*, 44(10), 1161–1165.

Sutton, R. E., & Boothroyd J. C. (1986). Evidence for trans-splicing in trypanosomes. *Cell*, 47, 527–535.

Sutton, W. S. (1903). The chromosomes in heredity. *Biological Bulletin*, 4, 231–251.

Suvà, M. L., Riggi, N., & Bernstein, B. E. (2013). Epigenetic reprogramming in cancer. *Science*, 339(6127), 1567–1570.

Tabery, J. (2014). *Beyond Versus: The Struggle to Understand the Interaction of Nature and Nurture*. Cambridge MA: MIT Press.

Takahashi, K., & Yamanaka, S. (2006). Induction of pluripotent stem cells from mouse embryonic and adult fibroblast cultures by defined factors. *Cell*, 126, 663–676.

Temin, H., & Mizutani, S. (1970). RNA-dependent DNA polymerase in virions of Rous sarcoma virus. *Nature*, 226, 1211–1213.

Thein, S. L. (2013). The molecular basis of β-thalassemia. *Cold Spring Harbor Perspectives in Medicine*, 3(5), a011700.

Thein, S. L., Old, J. M., Wainscoat, J. S. et al. (1984). Population and genetic studies suggest a single origin for the Indian deletion βo thalassaemia. *British Journal of Haematology*, 57, 271–278.

Thompson, W. C. (2013). Forensic DNA evidence: the myth of infallibility. In Krimsky, S., & Gruber, J. (Eds.) *Genetic Explanations: Sense and Nonsense*. Cambridge MA: Harvard University Press, 227–255.

Thöny, B., & Blau, N. (2006). Mutations in the BH4-metabolizing genes GTP cyclohydrolase I, 6-pyruvoyl-tetrahydropterin synthase, sepiapterin reductase, carbinolamine-4a-dehydratase, and dihydropteridine reductase. *Human Mutation*, 27(9), 870–878.

Tomasetti, C., & Vogelstein, B. (2015). Variation in cancer risk among tissues can be explained by the number of stem cell divisions. *Science*, 347(6217), 78–81.

Tschermak, E. (1950/1900). Concerning artificial crossing in *Pisum sativum*. *Genetics*, 35, 42–47.

Usifo, E., Leigh, S. E. A., Whittall, R. A., et al. (2012). Low-density lipoprotein receptor gene familial hypercholesterolemia variant database: update and pathological assessment. *Annals of Human Genetics*, 76, 387–401.

Varmus, H. (2006). The new era in cancer research. *Science*, 312(5777), 1162–1165.

Vassos, E., Collier, D. A., & Fazel, S. (2014). Systematic meta-analyses and field synopsis of genetic association studies of violence and aggression. *Molecular Psychiatry*, 19(4), 471–477.

Veeramachaneni, V., Makalowski, W., Galdzicki, M., Sood, R., & Makalowska, I. (2004). Mammalian overlapping genes: the comparative perspective. *Genome Research*, 14, 280–286.

Venter, J. C., Adams, M. D., Myers, E. W., et al. (2001). The sequence of the human genome. *Science*, 291(5507), 1304–1351.

Vettese-Dadey, M., Grant, P. A., Hebbes, T. R., et al. (1996). Acetylation of histone H4 plays a primary role in enhancing transcription factor binding to nucleosomal DNA in vitro. *EMBO Journal*, 15(10), 2508.

Vicedo, M. (1992). The human genome project: towards an analysis of the empirical, ethical, and conceptual issues involved. *Biology and Philosophy*, 7(3), 255–278.

Vickery, H. B. (1950). The origin of the word protein. *Yale Journal of Biology and Medicine*, 22(5), 387–393.

Visscher, P. M. (2008). Sizing up human height variation. *Nature Genetics*, 40(5), 489–490.

Visscher, P. M., Hill, W. G., & Wray, N. R. (2008). Heritability in the genomics era – concepts and misconceptions. *Nature Reviews Genetics*, 9(4), 255–266.

Visscher, P. M., Brown, M. A., McCarthy, M. I., & Yang, J. (2012). Five years of GWAS discovery. *American Journal of Human Genetics*, 90(1), 7–24.

Visscher, P. M., Medland, S. E., Ferreira M. A. R., et al. (2006). Assumption-free estimation of heritability from genome-wide identity-by-descent sharing between full siblings. *PLoS Genetics*, 2(3), e41. doi:10.1371/journal.pgen.0020041

Vogelstein, B., Papadopoulos, N., Velculescu, V. E., et al. (2013). Cancer genome landscapes. *Science*, 339(6127), 1546–1558.

Voit, E. O. (2016). *The Inner Workings of Life: Vignettes in Systems Biology*. Cambridge: Cambridge University Press.

von Meyenn, F., & Reik, W. (2015). Forget the parents: epigenetic reprogramming in human germ cells. *Cell*, 161(6), 1248–1251.

Waddington, C. H. (1942). The epigenotype. *Endeavour*, 1, 18–20 (reprinted in *International Journal of Epidemiology*, 41(1), 10–13).

Waddington, C. H. (1957). *The Strategy of the Genes: A Discussion of Some Aspects of Theoretical Biology*. London: Allen & Unwin.

Wain, H. M., Bruford, E. A., Lovering, R. C. et al. (2002). Guidelines for human gene nomenclature. *Genomics*, 79(4), 464–470.

Waller, J. C. (2001). Ideas of heredity, reproduction and eugenics in Britain, 1800–1875. *Studies in History and Philosophy of Science Part C: Studies in History and Philosophy of Biological and Biomedical Sciences*, 32(3), 457–489.

Walpole, B., Merson-Davies, A., & L. Dann (2011). *Biology for the IB Diploma Coursebook*. Cambridge: Cambridge University Press.

Walsh, D. (2006). Evolutionary essentialism. *British Journal for the Philosophy of Science* 57(2), 425–448.

Walsh, R., Thomson, K., Ware, J. S. et al. (2016). Reassessment of Mendelian gene pathogenicity using 7,855 cardiomyopathy cases and 60,706 reference samples. *Genetics in Medicine* (advance online publication).

Wang, E. T., Sandberg, R., Luo, S., et al. (2008). Alternative isoform regulation in human tissue transcriptomes. *Nature*, 456, 470–476.

Wang, X., Miller, D. C., Harman, R., Antczak, D. F., & Clark, A. G. (2013). Paternally expressed genes predominate in the placenta. *Proceedings of the National Academy of Sciences*, 110(26), 10705–10710.

Ward, W., McGonegal, R., Tostas, P., & Damon, A. (2008). *Pearson Baccalaureate: Higher level Biology for the IB diploma*. Harlow GB: Pearson Education Limited.

Waters, C. K. (1994). Genes made molecular. *Philosophy of Science*, 61, 163–85.

Waters, C. K. (2007). Causes that make a difference. *Journal of Philosophy*, 104, 551–579.

Watson, J. D. (1992) A personal view of the project. In Kevles, D. J. & Hood, L. (Eds.) *The Code of Codes: Scientific and Social Issues in the Human Genome Project.* Cambridge MA: Harvard University Press, 164–173.

Watson, J. D. (1996/1968). *The Double Helix: A Personal Account of the Discovery of the Structure of DNA.* New York: Touchstone.

Watson, J. D., & Crick, F. H. C. (1953a). Molecular structure of nucleic acids. *Nature,* 171, 737–738.

Watson, J. D., & Crick, F. H. C. (1953b). Genetical implications of the structure of deoxyribonucleic acid. *Nature,* 171, 964–967.

Weatherall, D. J. (2001). Phenotype – genotype relationships in monogenic disease: lessons from the thalassaemias. *Nature Reviews Genetics,* 2(4), 245–255.

Weatherall, D. J. (2004). The thalassemias: the role of molecular genetics in an evolving global health problem. *American Journal of Human Genetics,* 74(3), 385–392.

Weedon, M. N., Lango, H., Lindgren, C. M. et al. (2008). Genome-wide association analysis identifies 20 loci that influence adult height. *Nature Genetics,* 40(5), 575–583.

Weischenfeldt, J., Symmons, O., Spitz F., & Korbel J. O. (2013). Phenotypic impact of genomic structural variation: insights from and for human disease. *Nature Reviews Genetics,* 14, 125–138.

Weismann, A. (1893/1892). *The Germ-Plasm: A Theory of Heredity.* New York: Charles Scribner's Sons.

Weldon, W. F. R. (1902). Mendel's laws of alternative inheritance in peas. *Biometrika,* 1, 228–254.

Wheeler, D. A., Srinivasan, M., Egholm, M. et al. (2008). The complete genome of an individual by massively parallel DNA sequencing. *Nature,* 452(7189), 872–876.

Whipple, G. H., & Bradford, W. L. (1936). Mediterranean disease-thalassemia (erythroblastic anemia of Cooley): associated pigment abnormalities simulating hemochromatosis. *Journal of Pediatrics,* 9(3), 279–311.

Wilkins, J. S. (2009). *Species: A History of the Idea.* Berkeley CA: University of California Press.

Wilkins, J. S. (2013). Essentialism in biology. In Kampourakis, K. (Ed.) *The Philosophy of Biology: A Companion for Educators.* Dordrecht: Springer, 395–419.

Wilkins, M. H. F., Stokes A. R., & Wilson, H. R. (1953). Molecular structure of deoxypentose nucleic acids. *Nature,* 171, 738–740.

Wilson, E. B. (1896). *The Cell in Development and Inheritance.* London: Macmillan & Co. Ltd.

Winchester, A. M. (2013). The Work of Mendel. *Encyclopedia Britannica,* Retrieved February 20, 2014, www.britannica.com/EBchecked/topic/228936/genetics/261528/The-work-of-Mendel.

Witkin, H. A., Mednick, S., Schulsinger, F. et al. (1976). Criminality in XYY and XXY men: the elevated crime rate of XYY males is not related to aggression. It may be related to low intelligence. *Science,* 193(4253), 547–555.

Wolffe, A. P., & Pruss, D. (1996). Targeting chromatin disruption: transcription regulators that acetylate histones. *Cell,* 84(6), 817–819.

Wolpert, L. (1992). *The Unnatural Nature of Science: Why Science Does Not Make (Common) Sense*. Cambridge MA: Harvard University Press.

Wolpert, L. (2009). *How We Live and Why We Die: The Secret Lives of Cells*. London: Faber & Faber.

Wolpert, L. (2011). *Developmental Biology: A Very Short Introduction*. Oxford: Oxford University Press.

Wood, R. J., & Orel, V. (2005). Scientific breeding in Central Europe during the early nineteenth century: background to Mendel's later work. *Journal of the History of Biology*, 38, 239–272.

Wood, A. R., Esko, T., Yang, J., Vedantam, S., et al. (2014). Defining the role of common variation in the genomic and biological architecture of adult human height. *Nature Genetics*, 46(11), 1173–1186.

Wu, S., Powers, S., Zhu, W., & Hannun, Y. A. (2016). Substantial contribution of extrinsic risk factors to cancer development. *Nature*, 529(7584), 43–47.

Wu, C. S., Yu, C. Y., Chuang, C. Y. et al. (2014). Integrative transcriptome sequencing identifies trans-splicing events with important roles in human embryonic stem cell pluripotency. *Genome Research*, 24(1), 25–36.

Yamanaka, S. (2009). Elite and stochastic models for induced pluripotent stem cell generation. *Nature*, 460(7251), 49.

Yanai, I., & Lercher, M. (2016). *The Society of Genes*. Cambridge MA: Harvard University Press.

Yang, J., Benyamin, B., McEvoy, B. P. et al. (2010). Common SNPs explain a large proportion of the heritability for human height. *Nature Genetics*, 42(7), 565–569.

Yang, J., Manolio, T. A., Pasquale, L. R. et al. (2011). Genome partitioning of genetic variation for complex traits using common SNPs. *Nature Genetics*, 43(6), 519–525.

Yu, P., Ma, D., & Xu, M. (2005). Nested genes in the human genome. *Genomics*, 86, 414–422.

Zhang, D.-W., Lagace, T. A., Garuti, R. et al. (2007). Binding of proprotein convertase subtilisin/kexin type 9 to epidermal growth factor-like repeat A of low density lipoprotein receptor decreases receptor recycling and increases degradation. *Journal of Biological Chemistry*, 282, 18602–18612.

Zijlstra, M., Bix, M., Simister, N. E. et al. (1990). β2-microglobulin deficient mice lack CD4–8+ cytolytic T cells. *Nature*, 344, 742–746.

Glossary

acetylation The addition of an acetyl group (CH_3CO-) to an organic molecule such as a histone protein. The removal of the acetyl group is called deacetylation.

allele One of the several variants of a particular gene that "encodes" a particular protein or RNA molecule, and thus affects a particular biological process. Alleles are identified with particular parts of chromosomes, which are described as loci (singular locus).

allelic heterogeneity The phenomenon of multiple alleles in the same locus conferring the same phenotypic effect.

aneuploidy The presence of an abnormal number of chromosomes within a cell, usually one more or one less.

autosomes In species with two distinct sexes (sexual dimorphism) the chromosomes found in both sexes are called autosomes.

benign tumor An abnormal proliferation of cells. These cells are not invasive (i.e. they cannot penetrate the basement membrane lining them), and thus are not dangerous.

blastocyst An early mammal embryo of approximately 150 cells produced by cell divisions following fertilization. The blastocyst is a spherical cell mass consisting of an outer layer of cells (the trophoblast) and a cluster of cells in the interior (the inner cell mass).

character (biological) Any recognizable feature of an organism that can exist in a variety of character states and at several levels from the molecular to the organismal.

chromatin The complex of DNA and protein that composes chromosomes. Changes in chromatin structure are affected by DNA methylation and histone modifications. During cell division chromatin is further packed, forming chromosomes.

cis-regulatory elements DNA sequences that are implicated in the regulation of the expression of genes on the same chromosome.

complex traits Phenotypes considered to result from the independent action of many genes, environmental factors, and gene-environment interactions.

concepts Mental representations of the world, in terms of which our knowledge and understanding is formulated. Scientific concepts are systematic representations of the world through which explanations of and predictions about phenomena are possible.

crossing over The phenomenon of exchange of chromosome parts between two chromosomes during meiosis. This phenomenon results in new combinations of genes in offspring that did not exist in their parents.

development The processes of growth and differentiation that organisms undergo in their life cycle, e.g., from a fertilized egg to a sexually mature adult. It is also called ontogeny.

developmental (phenotypic) plasticity Modifiability of the phenotype during development. Individuals of the same species with the same genotype may exhibit phenotypic variation depending on local conditions.

DNA information It is often stated that DNA "encodes," "contains," "stores," or "transmits" information. DNA plays an important role in certain bioinformational relationships, usually as a message. It should be noted that bioinformation is not a property of DNA, but a complex relation in which DNA has an important role.

DNA methylation The addition of a methyl group to DNA at a cytosine that precedes a guanine.

dominance In classical genetics, a quality inherent in one of a pair of alleles in a diploid organism, the phenotype of which is manifest in the heterozygote.

driver gene mutation A mutation that directly or indirectly confers a selective growth advantage to the cell in which it occurs. Such mutations have an important role in cancer.

electropherogram The graphical output of electrophoresis devices in short tandem repeats (STRs) and sequencing analysis, showing fluorescence intensity as a function of molecular weight. In this case, the peaks at a particular wavelength (color) correspond to a specifically labeled molecule of a particular size.

embryo In humans, the organism that develops from the time of fertilization until the end of the eighth week of gestation, at which point it is called a fetus.

embryonic stem cells Pluripotent cells derived from the inner cell mass, which can differentiate in vitro into many different lineages and cell types, and, upon injection into blastocysts, can give rise to all tissues including the germline.

epigenetic reprogramming The process during which the epigenetic marks are erased and are created anew on a genome-wide scale. This happens at two distinct stages in mammals: in primordial germ cells when they have reached the embryonic gonads, and in the early embryo between the stage immediately after fertilization and the blastocyst stage.

epigenetics This term refers to all processes related to modifications above the level of DNA sequence that affect phenotypes.

epigenome The overall epigenetic state of a cell, i.e. which epigenetic marks (DNA methylation, histone methylation, or histone acetylation) it contains.

epistasis The phenomenon during which the effect of an allele at one locus may mask the effect of an allele at another locus.

essence The necessary properties of an entity that make it the kind of entity it is, often contrasted to "accidental" or contingent properties. These are the properties that all members of a kind must have, and the combination of which only members of this kind do, in fact, have.

eukaryote A macro- or microorganism that is not in Archaea or Bacteria, and which possesses discrete cellular compartments such as the nucleus.

exome The collection of exons in the human genome.

exons The DNA sequences that are retained in the mature mRNA after RNA splicing in eukaryotes. These are the DNA sequences that encode the information about which amino acids will be included in the protein.

expressivity The qualitatively different phenotypes that individuals with the same genotype can exhibit.

function The role of a component in the organization of a system.

functional explanations In biology, functional explanations answer the question of why particular organisms have a particular character by pointing out that the latter confers an advantage to those organisms because it efficiently performs a role.

gain of function When an allele causes higher levels of gene expression than the "normal" level.

gene A union of genomic sequences encoding a coherent set of potentially overlapping functional products.

gene function Refers to the contributions of genes to characters by, e.g., being implicated in the synthesis of a particular RNA or protein.

gene knockout The targeted disruption of particular genes in their actual biological context. It involves the replacement of the "normal" copy of a gene with an abnormal copy, which in turn results in the production of a non-functional protein or no protein at all.

gene structure Refers to features of genes (e.g. base sequence) that relate to the kind of molecules produced (e.g. a particular kind of RNA or protein), or how they "encode" some kind of information.

genetic code The specific correspondence between triplets of nucleotides and amino acids, on the basis of which a protein is synthesized after the translation of an mRNA.

genetic determinism The view that genes invariably determine characters, so that the outcomes are just a little, or not at all, affected by changes in the environment or by the different environments in which individuals live.

genetic essentialism The view that genes are fixed entities, which are transferred unchanged across generations and which are the essence of what we are by specifying characters from which their existence can be inferred.

genetic material Any nucleic acid with the propensity to be inherited and to interact with other cellular components as a source of sequence information, eventually affecting or being implicated in cellular processes with local or extended impact.

genetic reductionism The view that genes provide the ultimate explanation for characters, and so the best approach to explain these is by studying phenomena at the level of genes.

genome The totality of the DNA of an individual.

genome mosaicism The phenomenon of any individual being made up of a population of cells with differences in their genomes because of mutations.

genome-wide association (GWA) study An investigation of the association between common genetic variation and disease. This type of analysis requires a dense set of DNA markers that capture a substantial proportion of common variation across the genome and large numbers of participants.

genomic sequencing The methods and technologies used to determine the specific order of the bases A,T,C,G in a molecule of RNA or DNA.

genomics Biological research that focuses on whole genomes, i.e., the base sequence of the genetic material of organisms.

genotype The alleles related to a particular character that individuals carry.

germ cells The reproductive cells of the body. These and their whole lineage constitute the germline.

germline mutations Mutations that occur in any germ cell of the body, and which we usually inherit from our parents.

haploinsufficiency The phenomenon of an individual having only one functional copy of a gene rather than two, resulting in reduced levels of gene expression.

heredity The phenomenon of the transmission of some material substance that affects characters from ancestors to descendants.

heritability The proportion of the phenotypic variance in a population that is due to genotypic differences among individuals.

heterozygote An individual that carries two different alleles related to a particular character. A compound heterozygote is a heterozygote that carries two different pathogenic alleles and might thus exhibit the symptoms of the disease in the way that homozygotes do.

histone The main protein components of chromatin. The histones H2A, H2B, H3, and H4 form the nucleosome.

homologous recombination A procedure performed in order to insert an allele to a specific homologous genetic locus.

homozygote An individual that carries the same allele related to a particular character on both homologous chromosomes.

human embryonic stem cells Undifferentiated human cells that are derived from the inner cell mass of developing blastocysts and that are self-renewing, pluripotent, and capable of indefinitely dividing without differentiating in culture.

human pluripotent stem cells Human cells that have the capacity to differentiate into all tissues of an organism, but are not able to form embryonic components of the trophoblast and placenta and so cannot sustain the full development of an organism.

imprinting, genomic The epigenetic marking of a gene on the basis of parental origin, which in somatic tissues results in monoallelic expression, i.e., expression of one allele only.

in vitro Literally, "in glass", in a test tube, culture dish, or other artificial environment.

in vivo Within a living organism.

induced pluripotent stem cells Stem cells created by converting adult human cells into cells that are pluripotent and self-renewing.

inheritance The reception by offspring of material from parents. Inheritance was traditionally understood to involve property, but was later invoked to also explain how offspring have characteristics that resemble their parents' characteristics. Genetic inheritance refers to the process of transmission of genetic material across generations.

inner cell mass A cluster of cells attached to the inner wall of the blastocyst. In development, the inner cell mass gives rise to the organs and tissues of the organism.

linkage disequilibrium The nonrandom association of alleles at different loci in a population. SNPs that reside near one another on a chromosome are often inherited together because they are not separated by recombination events due to their proximity.

locus (plural loci) Particular parts of chromosomes in which particular alleles or genes are considered to be localized. The word "locus" is not a synonym for gene, but refers to a map position.

loss-of-function The phenomenon of an allele causing either partial or complete loss of gene expression.

malignant tumor An abnormal proliferation of cells driven by mutations in oncogenes or tumor suppressor genes that have already invaded their surrounding tissue.

meiosis Refers to the division of the nucleus in germ cells that results in the production of reproductive cells (sperm, ova). It literally means reduction and indicates that the number of chromosomes is reduced during cell division (e.g. human cells have forty-six chromosomes, whereas sperm and ova have each twenty-three chromosomes).

metastatic tumor A malignant tumor that has migrated away from its primary site to another tissue or organ.

methylation Addition of a methyl group to a protein, DNA, or other molecule.

missense mutation A single-nucleotide substitution (e.g. C to T) that results in an amino acid substitution (e.g. histidine to arginine).

mitosis Refers to the division of the nucleus in somatic cells. It stems from "mitos," literally meaning "filament," and indicates the filaments of chromatin observed in nuclei during cell division.

multipotency The ability of a cell to give rise to different cell types of a given cell lineage. These cells include most adult stem cells.

mutation A change in the structure of DNA, from a single nucleotide to long DNA segments that correspond to parts of chromosomes. Mutation does not necessarily have a "bad" outcome, although this is the case quite often, and it should be considered as synonymous to "change".

noncoding RNAs RNA molecules that do not encode proteins and that may have other regulatory functions.

nongenetic inheritance It refers to mechanisms by which the parental phenotypes affect the development of their offspring, and includes cellular epigenetic inheritance, which is the transmission from mother to daughter cell of variation in the molecular epigenetic regulation of gene expression, as well as behavioral interactions between parents and offspring.

nonsense mutation A single-nucleotide substitution (e.g. C to T) that results in the production of a stop codon.

nonsynonymous mutation A mutation that alters the encoded amino acid sequence of a protein. These include missense, nonsense, splice site, translation start, translation stop, and indel mutations.

nucleosome The basic unit of chromatin, containing DNA wrapped around a histone octamer (which is composed of two copies of each of histones H3, H4, H2A, and H2B).

oncogene A gene that, when activated by mutation, increases the selective growth advantage of the cell in which it resides.

passenger mutation A mutation that has no direct or indirect effect on the selective growth advantage of the cell in which it occurred.

penetrance It refers to the number of individuals with the same genotype who exhibit the same phenotype.

pharmacogenomics The selection of a drug and its dosage on the basis of an individual's genomic profile, with the aim to maximize the efficacy and minimize the toxicity of the drug.

phenotype The outcome of development with reference to a particular character.

plasticity, developmental, or phenotypic The potential of organisms with the same genotype to produce different phenotypes during development.

pleiotropy The phenomenon during which the effect of an allele at one locus affects multiple phenomena within the organism.

pluripotency The ability of a cell to give rise to all cells of the embryo. Cells of the inner cell mass of the blastocyst are pluripotent.

promoter A region within or near a gene that regulates its expression. RNA polymerases bind to promoters and start transcription.

purifying selection Selection against deleterious alleles in a population, which prevents their increase in frequency and which may cause their eventual elimination.

reaction norm The representation of the different phenotypes of a particular genotype in different environments.

recessiveness In classical genetics, a quality inherent in one of a pair of alleles in a diploid organism, the phenotype of which is not manifest in the heterozygote.

reductionism, methodological The approach to investigate and understand a biological system through its decomposition into the parts of which it consists and the study of their structure and function.

reductionism, ontological The commitment to the nature of the components of the system, and to the idea that any change in the system results only from changes in these components.

regulatory DNA sequences DNA sequences that are not transcribed to mRNA-like protein-coding sequences, but affect their expression. They act as switches that regulate protein synthesis, depending on the transcription factors that bind to them.

robustness, developmental The consistency of the phenotype of individuals irrespectively of the environment in which they live.

sex chromosomes The chromosomes that are not found in different numbers in the two sexes and that have loci implicated in sex development.

somatic mutations Mutations that occur in any cell of the body – except for the germline – after conception, such as those that initiate tumorigenesis.

stem cell A cell that can divide and produce a particular cell type.

structural variants Broadly defined, these are all variants that are not single nucleotide variants. They include insertions, deletions, block substitutions, inversions of DNA sequences, and copy-number differences.

systems biology A group of experimental, analytical, and modeling approaches, developed to explain how biological properties emerge through complex interactions.

totipotency The ability of a cell to give rise to all cells of an organism, including embryonic and extraembryonic tissues. Zygotes are totipotent.

transcriptome All of the RNAs transcribed from a genome.

translocation A specific type of rearrangement where regions from two non-homologous chromosomes are joined.

transposable elements (transposons) Sequences of DNA that can change their location within the genome of a single cell. During this process, called transposition, the sequences can cause mutations and change the organization of DNA in the genome.

tumor suppressor gene A gene that, when inactivated by mutation, increases the proliferation potential of the cell in which it is found.

uniparental disomy The condition that an organism inherits both chromosomes of a given pair from the same parent.

unipotency The capacity of a cell to sustain only one cell type or cell lineage. Examples are terminally differentiated cells, certain adult stem cells (testis stem cells), and committed progenitors (erythroblasts).

utility, clinical Whether a test can provide information about diagnosis, treatment, management, or prevention of a disease that will be helpful to people.

validity, analytical How well a test predicts the presence or absence of a particular gene or genetic change; in other words, whether the test can accurately detect whether a specific genetic variant is present or absent.

validity, clinical How well a genetic variant being analyzed is related to the presence, absence, or risk of a specific disease.

variants with unknown significance DNA variants for which there is no sufficient information to indicate whether or not they confer a pathogenic or protective effect.

Index

Printed in the United States
By Bookmasters